105 Algebra Problems
From the AwesomeMath Summer Program

105 Algebra Problems
From the AwesomeMath Summer Program

Titu Andreescu

Library of Congress Control Number: 2012920776

ISBN-13: 978-0-9799269-5-2 ISBN-10: 0-9799269-5-5

© 2013 XYZ Press, LLC

All rights reserved. This work may not be translated or copied in whole or in part without the written permission of the publisher (XYZ Press, LLC, 3425 Neiman Rd., Plano, TX 75025, USA) and the authors except for brief excerpts in connection with reviews or scholarly analysis. Use in connection with any form of information storage and retrieval, electronic adaptation, computer software, or by similar or dissimilar methodology now known or hereafter developed is forbidden. The use in this publication of tradenames, trademarks, service marks and similar terms, even if they are not identified as such, is not to be taken as an expression of opinion as to whether or not they are subject to proprietary rights.

9 8 7 6 5 4 3 2 1

www.awesomemath.org

Cover design by Iury Ulzutuev

Contents

1	Preface	1
2	Completing the square and quadratic equations	2
3	Factorizations and algebraic identities	14
4	Factoring expressions involving $a-b$, $b-c$, $c-a$	27
5	Factoring $a^3 + b^3 + c^3 - 3abc$	36
6	AM-GM and Hölder's inequality	40
7	Lagrange's identity and the Cauchy-Schwarz inequality	52
8	Making linear combinations	65
9	Fixed points and monotonicity	77
10	The floor function	83
11	Taking advantage of symmetry	95
12	Introductory problems	111
13	Advanced problems	117
14	Solutions to introductory problems	123
15	Solutions to advanced problems	157
16	Other Books from XYZ Press	200

1 Preface

The main purpose of this book is to provide an introduction to central topics in elementary Algebra from a problem-solving point of view. While working with students who were preparing for various mathematics competitions or exams, I observed that fundamental algebraic techniques were not part of their mathematical repertoire. Since algebraic skills are not only critical to Algebra itself but also to numerous other mathematical fields, a lack of such knowledge can drastically hinder a student's performance. Taking the above observations into account, I put together this introductory book using both simple and challenging examples which shed light upon essential algebraic strategies and techniques, as well as their application in diverse meaningful problems. This work is the first volume in a series of such books.

Regarding the structure of the book, the featured topics are elementary and classical, including factorizations, algebraic identities, inequalities, algebraic equations and systems of equations. More advanced concepts such as complex numbers, exponents and logarithms, as well as other topics are generally avoided. Nevertheless, some problems are constructed using properties of complex numbers which challenge and expose the reader to a broader spectrum of mathematics. Each chapter focuses on specific methods or strategies and provides an ample collection of accompanying problems that graduate in difficulty and complexity. In order to assist the reader with verifying mastery of the theoretical component, I included 105 problems in the last sections of the book, of which 52 are introductory and 53 advanced. All problems come together with solutions, many employing several approaches and providing the motivation behind the solutions offered.

Enjoy the problems!

2 Completing the square and quadratic equations

While the identity
$$(a+b)^2 = a^2 + 2ab + b^2$$
is easy to check (simply expand the left hand-side, by writing it as $(a+b)(a+b)$), things are a little bit more difficult in real life. Indeed, most of the time we have to do exactly the opposite: we are given a quadratic expression and we want to express it as a sum of squares (or some linear combination of squares). The idea is quite simple: we fix one variable, say x of that expression, such that the expression becomes a quadratic polynomial in x. Say our expression is ax^2+bx+c for some real numbers a, b, c (which may themselves be complicated expressions depending on other real numbers!). Then we complete the square by writing
$$ax^2 + bx + c = a\left(x^2 + \frac{b}{a}x + \frac{c}{a}\right) =$$
$$a\left(\left(x + \frac{b}{2a}\right)^2 - \frac{b^2}{4a^2} + \frac{c}{a}\right) = a\left(x + \frac{b}{2a}\right)^2 - \frac{\Delta}{4a},$$
where $\Delta = b^2 - 4ac$ is the discriminant of the expression ax^2+bx+c. This eliminates the variable x, by including all of its appearances in the term $a\left(x + \frac{b}{2a}\right)^2$. Now, $-\frac{\Delta}{4a}$ may (or may not...) itself be some quadratic expression in different variables, so we can apply the same reasoning to write it as a sum of squares.

In particular, the previous discussion applies to the quadratic equation
$$ax^2 + bx + c = 0,$$
where a, b, c are given real numbers, with a nonzero (if $a = 0$, then we obtain a linear equation). The previous paragraph shows that the equation can be written as
$$\left(x + \frac{b}{2a}\right)^2 = \frac{\Delta}{4a^2}.$$
If the equation has real solutions, then the left hand-side must be nonnegative (as is the square of any real number). Hence so must be the right hand-side, which means that $\Delta \geq 0$. In this case, we can solve the previous equation by taking square roots, and we end up with the solutions
$$x_1 = \frac{-b + \sqrt{\Delta}}{2a}, \quad x_2 = \frac{-b - \sqrt{\Delta}}{2a},$$
which are equal if and only if $\Delta = 0$. Hence we can summarize our discussion in:

Theorem 2.1. *Let a, b, c be real numbers with $a \neq 0$ and let*

$$\Delta = b^2 - 4ac.$$

Then the quadratic equation

$$ax^2 + bx + c = 0$$

has either:
- *no real solution, if $\Delta < 0$.*
- *exactly one real solution, if $\Delta = 0$.*
- *two real solutions if $\Delta > 0$.*

Note that the previous discussion also gives a nice way of solving quadratic inequalities or proving inequalities involving quadratic expressions: since

$$ax^2 + bx + c = a\left(\left(x + \frac{b}{2a}\right)^2 + \frac{-\Delta}{4a^2}\right),$$

we see that the expression $ax^2 + bx + c$ has constant sign (equal to that of a) when $\Delta \leq 0$. On the other hand, if $\Delta > 0$ and if $x_1 \leq x_2$ are the real solutions of the equation $ax^2 + bx + c = 0$, then the inequality $ax^2 + bx + c \leq 0$ is equivalent to $a(x - x_1)(x - x_2) \leq 0$. If $a > 0$, this is in turn equivalent to $x \in [x_1, x_2]$, while if $a < 0$, this is equivalent to $x \notin (x_1, x_2)$. To summarize, suppose for the sake of simplicity that $a > 0$. Then
- if $\Delta = b^2 - 4ac < 0$, then $ax^2 + bx + c > 0$ for all real numbers x.
- If $\Delta = 0$, then $ax^2 + bx + c \geq 0$ for all real numbers x, with equality if and only if $x = -\frac{b}{2a}$.
- If $\Delta > 0$, then the equation $ax^2 + bx + c = 0$ has two real roots, say $x_1 < x_2$, and we have $ax^2 + bx + c < 0$ if and only if $x \in (x_1, x_2)$.

A consequence of this discussion is the following important fact (which is a very special case of a general theorem in real analysis):

Theorem 2.2. *Let $f(x) = ax^2 + bx + c$ be a quadratic polynomial and let $u \leq v$ be real numbers such that $f(u)f(v) < 0$. Then the equation $f(x) = 0$ has at least one solution in (u, v).*

Proof. Since f changes sign between u and v, its discriminant $\Delta = b^2 - 4ac$ must be positive and so the equation $f(x) = 0$ has two distinct solutions $x_1 < x_2$. If none of them belongs to (u, v), then the previous discussion shows that either $f(u), f(v) > 0$ or $f(u), f(v) < 0$ (according to the sign of a). But this contradicts the hypothesis that $f(u)$ and $f(v)$ have different signs. □

Actually the previous theorem holds for any polynomial function (and more generally for continuous functions), but the proof is beyond the scope of this introductory book. Another very important result concerning quadratic equations is the following:

Theorem 2.3. *(Vieta's relations for quadratic equations)* Let a, b, c be real numbers, with $a \neq 0$ and let x_1, x_2 be the roots of the equation $ax^2 + bx + c = 0$. Then
$$x_1 + x_2 = -\frac{b}{a} \quad \text{and} \quad x_1 x_2 = \frac{c}{a}.$$

Proof. Since $ax^2 + bx + c = 0$ has roots x_1, x_2, we must have an equality of polynomials
$$ax^2 + bx + c = a(x - x_1)(x - x_2) = ax^2 - a(x_1 + x_2)x + ax_1 x_2.$$

Identifying coefficients yields
$$x_1 + x_2 = -\frac{b}{a} \quad \text{and} \quad x_1 x_2 = \frac{c}{a},$$

which is exactly what we wanted to prove. $\square$

We remark that the previous theorem holds for complex roots, and roots with multiplicity, with the same proof.

It is now time for practice: we will see how the above theoretical facts really apply in practice.

Example 2.1. Solve the equation
$$\frac{(2x-1)^2}{2} + \frac{(3x-1)^2}{3} + \frac{(6x-1)^2}{6} = 1.$$

Solution. Expanding each term and collecting terms according to the successive powers of x yields the following equivalent equations
$$2x^2 - 2x + \frac{1}{2} + 3x^2 - 2x + \frac{1}{3} + 6x^2 - 2x + \frac{1}{6} = 1,$$

$$11x^2 - 6x = 0 \quad \text{or} \quad x(11x - 6) = 0.$$

Hence the solutions are $x = 0$ and $x = \frac{6}{11}$.

Example 2.2. Find the greatest integer n for which the equation
$$\frac{1}{x-1} - \frac{1}{nx} + \frac{1}{x+1} = 0$$

has real solutions.

Solution. Write the equation successively as

$$\frac{1}{x-1} + \frac{1}{x+1} = \frac{1}{nx},$$

then $\frac{2x}{x^2-1} = \frac{1}{nx}$, $2nx^2 = x^2 - 1$ and finally $(-2n+1)x^2 = 1$. Since $x^2 \geq 0$ for all real numbers x, we deduce that if the equation has real solutions, then $-2n + 1 > 0$, hence $n \leq 0$. We cannot have $n = 0$, since then $\frac{1}{nx}$ wouldn't make sense. Hence the largest n is at most -1. And indeed $n = -1$ gives the real solutions $x = \frac{1}{\sqrt{3}}$ and $x = -\frac{1}{\sqrt{3}}$, so the answer is $n = -1$.

Example 2.3. Solve the system of equations

$$\begin{cases} x - y = 3 \\ x^2 + (x+1)^2 = y^2 + (y+1)^2 + (y+2)^2. \end{cases}$$

Solution. We take advantage of the fact that the first equation is very simple and express $x = y + 3$, replacing this value in the second equation. This yields the quadratic equation

$$(y+3)^2 + (y+4)^2 = y^2 + (y+1)^2 + (y+2)^2.$$

Expanding each term and collecting similar terms, we obtain the equivalent equation

$$2y^2 + 14y + 25 = 3y^2 + 6y + 5 \quad \text{or} \quad y^2 - 8y - 20 = 0.$$

Solving this equation gives $y = -2, 10$ and since $x = y + 3$, we obtain the solutions $(x, y) = (1, -2)$ and $(13, 10)$.

Example 2.4. Evaluate

$$\frac{1}{\sqrt{x + 2\sqrt{x-1}}} + \frac{1}{\sqrt{x - 2\sqrt{x-1}}},$$

where $1 \leq x < 2$.

Solution. We start by simplifying each fraction, by completing the squares at the denominator. We have

$$x + 2\sqrt{x-1} = x - 1 + 2\sqrt{x-1} + 1 = (\sqrt{x-1} + 1)^2$$

and

$$x - 2\sqrt{x-1} = x - 1 - 2\sqrt{x-1} + 1 = (\sqrt{x-1} - 1)^2.$$

Thus, paying attention to the fact that $\sqrt{a^2} = |a|$ and $\sqrt{x-1} - 1 < 0$ (since $x < 2$), we obtain

$$\frac{1}{\sqrt{x+2\sqrt{x-1}}} + \frac{1}{\sqrt{x-2\sqrt{x-1}}} = \frac{1}{1+\sqrt{x-1}} + \frac{1}{1-\sqrt{x-1}}$$

$$= \frac{2}{(1+\sqrt{x-1})(1-\sqrt{x-1})} = \frac{2}{1-(x-1)} = \frac{2}{2-x}.$$

Example 2.5. Solve the system of equations

$$\begin{cases} x + \frac{1}{y} = -1 \\ y + \frac{1}{z} = \frac{1}{2} \\ z + \frac{1}{x} = 2. \end{cases}$$

Solution. The idea is quite simple: we express everything in terms of one variable. Namely, from the first equation we can express x in terms of y, obtaining $x = -1 - \frac{1}{y}$. The second equation gives

$$z = \frac{2}{1-2y}.$$

Replacing these values in the last equation, we obtain

$$\frac{2}{1-2y} - \frac{y}{1+y} = 2.$$

Clearing denominators and simplifying the resulting equation, we arrive at

$$y + 2y^2 = 0.$$

Note that $y \neq 0$, since otherwise $\frac{1}{y}$ wouldn't make sense. We conclude that $y = -\frac{1}{2}$. Coming back to $x = -1 - \frac{1}{y}$ and $z = \frac{2}{1-2y}$, we obtain $x = z = 1$, hence the system has the unique solution $(1, -\frac{1}{2}, 1)$.

Example 2.6. Solve the equation

$$\frac{1}{3x-1} + \frac{1}{4x-1} + \frac{1}{7x-1} = 1.$$

Solution. The algebra would be quite nasty if we tried to clear denominators. Instead, we rewrite the equation as

$$\frac{1}{3x-1} + \frac{1}{4x-1} = 1 - \frac{1}{7x-1}$$

or equivalently
$$\frac{4x-1+3x-1}{(3x-1)(4x-1)} = \frac{7x-1-1}{7x-1}.$$
We remark the common factor $7x-2$, which already gives us the solution $x = \frac{2}{7}$. Suppose that $x \neq \frac{2}{7}$ is another solution. Then dividing the previous relation by $7x-2$ yields
$$\frac{1}{(3x-1)(4x-1)} = \frac{1}{7x-1} \quad \text{or} \quad 12x^2 - 7x + 1 = 7x - 1.$$
This can be further simplified to $6x^2 - 7x + 1 = 0$. Solving this quadratic equation yields the other solutions $x = 1, \frac{1}{6}$ of the equation. Hence the equation has three solutions, given by $\frac{2}{7}, 1, \frac{1}{6}$.

Example 2.7. Find all pairs (a, b) of positive real numbers such that
$$4a + 9b = \frac{9}{a} + \frac{4}{b} = 12.$$

Solution. We write the second equation as
$$\frac{9b + 4a}{ab} = 12$$
and we observe that the numerator equals 12 by hypothesis. Thus $ab = 1$, that is $b = \frac{1}{a}$. Replacing this value of b in the equation $4a + 9b = 12$ we obtain $4a + \frac{9}{a} = 12$. Clearing denominators, we obtain a quadratic equation $4a^2 - 12a + 9 = 0$, which has the unique solution $a = \frac{3}{2}$. Going back to the system, we obtain $b = \frac{2}{3}$.

If you found the first step (establishing that $ab = 1$) tricky, we can work more directly as follows: from the equation $4a + 9b = 12$ we express b in terms of a. We replace this value of b in the equation $\frac{9}{a} + \frac{4}{b} = 12$, obtaining a quadratic equation in a, with the unique solution $a = \frac{3}{2}$.

Example 2.8. If a is a real number such that $a - \frac{1}{a} = 1$, find $a^4 + \frac{1}{a^4}$.

Solution. It is easier to realize what you shouldn't do in this exercise: you should not solve the equation $a - \frac{1}{a} = 1$ and then plug in the values you get to compute $a^4 + \frac{1}{a^4}$ (of course, with a lot of nasty computations one would obtain the desired answer, but this is far from being an elegant approach). Let us take the square of the given relation $a - \frac{1}{a} = 1$, and obtain
$$a^2 + \frac{1}{a^2} - 2 = 1, \quad \text{that is} \quad a^2 + \frac{1}{a^2} = 3.$$
Now, all we have to do is to repeat the process: we take the square of the last relation and obtain
$$a^4 + \frac{1}{a^4} + 2 = 9, \quad \text{hence} \quad a^4 + \frac{1}{a^4} = 9 - 2 = 7.$$

Example 2.9. Solve the equation

$$x^4 - 97x^3 + 2012x^2 - 97x + 1 = 0.$$

Solution. The key point is that the equation is symmetric. Dividing by x^2, we obtain the equivalent equation

$$x^2 - 97x + 2012 - \frac{97}{x} + \frac{1}{x^2} = 0$$

We reduce this to a quadratic equation by setting

$$x + \frac{1}{x} = y.$$

Then $x^2 + \frac{1}{x^2} + 2 = y^2$, hence the previous equation becomes

$$y^2 - 97y + 2010 = 0, \quad \text{or} \quad (y-30)(y-67) = 0.$$

Thus $y = 30$ or $y = 67$. Now, remember that $y = x + \frac{1}{x}$, hence we obtain the quadratic equation $x^2 - xy + 1 = 0$. Solving the two equations $x^2 - 30x + 1 = 0$ and $x^2 - 67x + 1 = 0$ gives the solutions

$$x = \frac{67 \pm \sqrt{4485}}{2}, \quad \text{and} \quad x = 15 \pm \sqrt{224}.$$

Example 2.10. Let a, b, c be real numbers such that $a \geq b \geq c$. Prove that

$$a^2 + ac + c^2 \geq 3b(a - b + c).$$

Solution. Let us rewrite the inequality as

$$3b^2 - 3b(a+c) + a^2 + ac + c^2 \geq 0.$$

This is a quadratic inequality in b. The discriminant is

$$9(a+c)^2 - 12(a^2 + ac + c^2) = -3(a^2 - 2ac + c^2) = -3(a-c)^2 \leq 0,$$

thus the polynomial $3x^2 - 2x(a+c) + a^2 + ac + c^2$ takes only nonnegative values (its leading coefficient 3 is positive). In particular, its value at b is nonnegative, and the result follows.

We can also try to complete squares, writing the inequality as

$$12b^2 - 12b(a+c) + 4(a^2 + ac + c^2) \geq 0,$$

then

$$3(2b - (a+c))^2 + (a-c)^2 \geq 0.$$

Example 2.11. Prove that $3(x+y+1)^2 + 1 \geq 3xy$ for all $x, y \in \mathbf{R}$.

Solution. Let us write $x + y = a$ and $xy = b$. We need to prove that $3(a+1)^2 + 1 \geq 3b$. Now, the equation $t^2 - at + b = 0$ has the real solutions x, y, hence its discriminant is nonnegative, that is $a^2 \geq 4b$ (of course, we can also give a direct proof, since the inequality is equivalent to $(x-y)^2 \geq 0$). Thus $b \leq \frac{a^2}{4}$ and it suffices to prove that

$$3(a+1)^2 + 1 \geq \frac{3}{4}a^2.$$

Multiplying by 4, expanding $(a+1)^2$ and rearranging terms reduces the inequality to

$$9a^2 + 24a + 16 \geq 0,$$

equivalent to $(3a+4)^2 \geq 0$, thus true.

We note that an alternative solution consists in completing the square, which allows us to rewrite the inequality as

$$3\left(x + \frac{1}{2}y + 1\right)^2 + \left(\frac{3}{2}y + 1\right)^2 \geq 0.$$

Example 2.12. Find all pairs (x, y) of real numbers such that

$$4x^2 + 9y^2 + 1 = 12(x + y - 1).$$

Solution. Let us separate the variables by writing the equation in the form

$$4x^2 - 12x + 9y^2 - 12y + 13 = 0.$$

Next, we complete the square to obtain

$$(2x - 3)^2 + (3y - 2)^2 = 0.$$

Since a sum of squares equals zero if and only if each square is zero, it follows that $2x - 3 = 0$ and $3y - 2 = 0$. Thus there is only one solution, given by $x = \frac{3}{2}$ and $y = \frac{2}{3}$.

Example 2.13. Prove that if $a \geq b > 0$, then

$$\frac{(a-b)^2}{8a} \leq \frac{a+b}{2} - \sqrt{ab} \leq \frac{(a-b)^2}{8b}.$$

Solution. We complete the square to obtain

$$\frac{a+b}{2} - \sqrt{ab} = \frac{a - 2\sqrt{a}\sqrt{b} + b}{2} = \frac{(\sqrt{a} - \sqrt{b})^2}{2}.$$

On the other hand, we have
$$a - b = \sqrt{a}^2 - \sqrt{b}^2 = (\sqrt{a} - \sqrt{b})(\sqrt{a} + \sqrt{b}).$$
Dividing by $(\sqrt{a} - \sqrt{b})^2$, we are therefore reduced to proving the inequalities
$$\frac{(\sqrt{a} + \sqrt{b})^2}{8a} \leq \frac{1}{2} \leq \frac{(\sqrt{a} + \sqrt{b})^2}{8b}.$$
The inequality on the left is equivalent (after multiplication by $8a$ and taking square roots) to
$$\sqrt{a} + \sqrt{b} \leq 2\sqrt{a}$$
and is an immediate consequence of $a \geq b$. We proceed similarly with the inequality on the right.

Example 2.14. Simplify the expression
$$\frac{4}{4x^2 + 12x + 9} - \frac{12}{6x^2 + 5x - 6} + \frac{9}{9x^2 - 12x + 4}.$$

Solution. Here it is easier to say what you should **not** do: clear denominators! Indeed, that would give a terrible mess and chances to solve the exercises with this approach are close to zero. Instead, let us analyze a little bit each term in the sum, more precisely its denominator. Each denominator is a quadratic polynomial in x, so a natural approach would be to see whether it can itself be factored. Of course, a sum of products is not something very enlightening, but one might hope that the denominators have a common factor. Well, solving the quadratic equations $4x^2 + 12x + 9 = 0$, $6x^2 + 5x - 6 = 0$ and $9x^2 - 12x + 4 = 0$, or by completing the square we end up with
$$4x^2 + 12x + 9 = (2x + 3)^2,$$
$$6x^2 + 5x - 6 = (2x + 3)(3x - 2)$$
$$9x^2 - 12x + 4 = (3x - 2)^2.$$
It turns out that the don't have a common factor, but they have a quite good shape: if $a = 2x + 3$ and $b = 3x - 2$, then our expression is simply
$$\frac{4}{a^2} - \frac{12}{ab} + \frac{9}{b^2} = \frac{4b^2 - 12ab + 9a^2}{(ab)^2}.$$
Again, it is easy to recognize that the numerator is the square of $(2b - 3a)^2$. Since
$$2b - 3a = 2(3x - 2) - 3(2x + 3) = -13,$$
we obtain the nice equality
$$\frac{4}{4x^2 + 12x + 9} - \frac{12}{6x^2 + 5x - 6} + \frac{9}{9x^2 - 12x + 4} = \left(\frac{13}{6x^2 + 5x - 6}\right)^2.$$

Example 2.15. Solve in real numbers the equation

$$x^4 + 16x - 12 = 0.$$

Solution. We will try to find a, b, c such that the left hand-side can be written as $(x^2 + a)^2 - (bx + c)^2$. If we can find such numbers a, b, c, then solving the equation will come down to solving two quadratic equations $x^2 + a = bx + c$ and $x^2 + a + bx + c = 0$.

The identity

$$x^4 + 16x - 12 = (x^2 + a)^2 - (bx + c)^2$$

is equivalent to the chain of equalities

$$2a = b^2, \quad 16 = -2bc, \quad a^2 - c^2 = -12.$$

Thus $a = \frac{b^2}{2}$, $c = -\frac{8}{b}$ and replacing these in the last equation yields

$$\frac{b^4}{4} - \frac{64}{b^2} = -12.$$

Let $b^2 = 4d$. The equation becomes $4d^2 - \frac{16}{d} = -12$ and we easily recognize the root $d = 1$. Thus we can take $b = 2$ and then $a = \frac{b^2}{2} = 2$ and $c = -\frac{8}{b} = -4$.

Now, it remains to solve the equations $x^2 + 2 = 2x - 4$ and $x^2 + 2 = -2x + 4$. The first one has no real solutions since it can be written as $(x-1)^2 + 5 = 0$, while the second one can be written $(x+1)^2 = 3$ and has the solutions

$$x_1 = -1 - \sqrt{3}, \quad x_2 = \sqrt{3} - 1.$$

Example 2.16. The equation $x^4 - 4x = 1$ has two real roots. Find their product.

Solution. Let us add $2x^2 + 1$ to both terms, in order to complete the square in the left hand-side. We obtain the equivalent equation

$$(x^2 + 1)^2 = 2x^2 + 4x + 2 = 2(x+1)^2.$$

This is equivalent to $x^2 + 1 = \sqrt{2}(x+1)$ or $x^2 + 1 = -\sqrt{2}(x+1)$. The first equation is $x^2 - \sqrt{2}x + 1 - \sqrt{2} = 0$ and its discriminant is $\Delta = 4\sqrt{2} - 2 > 0$. Hence it has two solutions x_1, x_2 and their product is given by Vieta's formulae: $x_1 x_2 = 1 - \sqrt{2}$. Since we already know that the initial equation has two real roots, they must be x_1, x_2 and we are done. Of course, it would be very easy to check that the equation $x^2 + 1 = -\sqrt{2}(x+1)$ has no real root, since its discriminant is negative. Thus the answer of the problem is $1 - \sqrt{2}$.

We can also solve this as follows we look for a, b, c such that
$$x^4 - 4x - 1 = (x^2 + a)^2 - (bx + c)^2$$
for all x, which is equivalent to
$$2a = b^2, \quad bc = 2, \quad a^2 - c^2 = -1.$$
Replacing $a = \frac{b^2}{2}$ and $c = \frac{2}{b}$ in the last equation yields
$$\frac{b^4}{4} - \frac{4}{b^2} = -1.$$

Setting $b^2 = 4d$, this gives us the third degree equation $4d^3 + d - 1 = 0$, with the apparent solution $d = \frac{1}{2}$. Thus $b^2 = 2$ and we can take $b = \sqrt{2}$, then $a = 1$ and $c = \sqrt{2}$. The original equation is therefore reduced to the resolution of the equations $x^2 + 1 = \sqrt{2}x + \sqrt{2}$ and $x^2 + 1 = -\sqrt{2}x - \sqrt{2}$. As above, we obtain the product of the real roots equals $1 - \sqrt{2}$.

Example 2.17. Solve in real numbers the equation
$$(x+1)(x+2)(x+3)(x+4) = 360.$$

Solution. Expanding the product is out of the question, so there is certainly some trick here. We try to pair the factors in the product defining the left hand-side. If we pair the first two and then the last two we obtain the equation $(x^2+3x+2)(x^2+7x+12) = 360$, which is not simpler. The same thing happens if we pair the first and the third factor, but a miracle happens if we pair the first and the last factor: we obtain the equation
$$(x^2 + 5x + 4)(x^2 + 5x + 6) = 360,$$
which is fourth degree equation in x, but a **quadratic** equation in $y = x^2 + 5x$. And while solving quartic equations is hard, solving quadratic ones is straightforward. Namely, the equation $(y+4)(y+6) = 360$ is equivalent to $y^2 + 10y - 336 = 0$, with solutions $y = 14$ and $y = -24$. Next, we have to solve the equations $x^2 + 5x = 14$ and $x^2 + 5x = -24$. The discriminant of the second one is negative, so it does not give real solutions. On the other hand, the equation $x^2 + 5x = 14$ has solutions $x = -7$ and $x = 2$. Hence these two numbers are the solutions of the initial equation.

Example 2.18. Find all $n > 1$ such that
$$x_1^2 + x_2^2 + \ldots + x_n^2 \geq x_n(x_1 + x_2 + \ldots + x_{n-1})$$
for all real numbers $x_1, \ldots, x_n$.

Solution. We write the inequality as
$$x_1^2 - x_1 x_n + x_2^2 - x_2 x_n + \ldots + x_{n-1}^2 - x_{n-1} x_n + x_n^2 \geq 0.$$

We complete the squares to get the equivalent inequality
$$\left(x_1 - \frac{x_n}{2}\right)^2 + \ldots + \left(x_{n-1} - \frac{x_n}{2}\right)^2 - \frac{n-1}{4} x_n^2 + x_n^2 \geq 0,$$

that is
$$\left(x_1 - \frac{x_n}{2}\right)^2 + \ldots + \left(x_{n-1} - \frac{x_n}{2}\right)^2 \geq \frac{n-5}{4} x_n^2.$$

If $n \leq 5$, then the right hand-side is nonpositive and the left hand-side is nonnegative, hence the inequality holds. On the other hand, if $n > 5$, then we can choose
$$x_1 = \frac{x_n}{2}, \ldots, x_{n-1} = \frac{x_n}{2}, \quad x_n = 1$$
and the inequality is no longer true. Hence the answer is $n = 2, 3, 4, 5$.

3 Factorizations and algebraic identities

There are a few classical algebraic identities that play a crucial role in almost all branches of mathematics. In this section we recall some of them and give many examples of applications to factorization of algebraic expressions. Being able to recognize factorizations of (sometimes complicated) algebraic expressions is fundamental, since this often plays an important role in solving equations, systems of equations or proving inequalities.

A first fundamental identity is

$$a^2 - b^2 = (a-b)(a+b).$$

This holds for all real numbers a, b (and is actually much more general than that, but we will stick to real numbers from now on) and follows easily by expanding the right hand-side and canceling the terms ab and $-ba$. Though very simple, this identity is crucial when factoring or simplifying more complicated algebraic expressions. It is also a special case of a more general, and also very handy identity, which describes a partial factorization of $a^n - b^n$. Note that $a^n - b^n$ vanishes when $a = b$, hence it must have a factor of $a - b$. We actually have

$$a^n - b^n = (a-b)(a^{n-1} + a^{n-2}b + ... + ab^{n-2} + b^{n-1})$$

for all real numbers a, b and all positive integers n. Indeed, we have

$$(a-b)(a^{n-1} + a^{n-2}b + ... + ab^{n-2} + b^{n-1}) =$$

$$a(a^{n-1} + a^{n-2}b + ... + b^{n-1}) - b(a^{n-1} + ... + b^{n-1}) =$$

$$a^n + a^{n-1}b + ... + ab^{n-1} - a^{n-1}b - ... - ab^{n-1} - b^n = a^n - b^n,$$

by canceling equal terms. For instance, if $n = 3$ we get the very useful identity

$$a^3 - b^3 = (a-b)(a^2 + ab + b^2)$$

and for $n = 4$ we obtain

$$a^4 - b^4 = (a-b)(a^3 + a^2b + ab^2 + b^3).$$

One may wonder if we can still factor $a^2 + ab + b^2$ and $a^3 + a^2b + ab^2 + b^3$. This is not the case for $a^2 + ab + b^2$, but the answer is positive for $a^3 + a^2b + ab^2 + b^3$, since

$$a^3 + a^2b + ab^2 + b^3 = a^2(a+b) + b^2(a+b) = (a+b)(a^2 + b^2).$$

We obtain therefore

$$a^4 - b^4 = (a-b)(a+b)(a^2 + b^2).$$

We could have also obtained this directly from the identity

$$x^2 - y^2 = (x - y)(x + y),$$

by applying it twice:

$$a^4 - b^4 = (a^2)^2 - (b^2)^2 = (a^2 - b^2)(a^2 + b^2) = (a - b)(a + b)(a^2 + b^2).$$

A similar argument yields the nice identities

$$a^{2^n} - b^{2^n} = (a - b)(a + b)(a^2 + b^2)(a^4 + b^4)...(a^{2^{n-1}} + b^{2^{n-1}})$$

and

$$a^{3^n} - b^{3^n} = (a-b)(a^2+ab+b^2)(a^6+a^3b^3+b^6)...(a^{2\cdot 3^{n-1}}+a^{3^{n-1}}b^{3^{n-1}}+b^{2\cdot 3^{n-1}}).$$

Before going on to explicit examples, we insist on the fact that one should know the following identities, which are all special cases of the binomial formula. However, it is very important to contemplate a while their coefficients, since this is how we recognize them in practice (it is not enough to know that there is a general formula-here the binomial formula-giving rise to all these identities, it is much more important to **see** which identity you need when solving a specific problem): for all real numbers a, b we have

$$(a + b)^2 = a^2 + 2ab + b^2, \quad (a + b)^3 = a^3 + 3a^2b + 3ab^2 + b^3,$$

$$(a + b)^4 = a^4 + 4a^3b + 6a^2b^2 + 4ab^3 + b^4,$$

$$(a + b)^5 = a^5 + 5a^4b + 10a^3b^2 + 10a^2b^3 + 5ab^4 + b^5.$$

Note that we can also write some of these as

$$(a + b)^3 = a^3 + b^3 + 3ab(a + b),$$

$$(a + b)^5 = a^5 + b^5 + 5ab(a^3 + b^3) + 10a^2b^2(a + b).$$

Finally, another classical and very useful identity is

$$(a + b + c)^2 = a^2 + b^2 + c^2 + 2(ab + bc + ca),$$

which follows from

$$(a + b + c)^2 = a^2 + 2a(b + c) + (b + c)^2 =$$

$$a^2 + 2ab + 2ac + b^2 + c^2 + 2bc = a^2 + b^2 + c^2 + 2(ab + bc + ca).$$

Let us now see how these things really work in practice. Keep in mind that quite often one must combine the previous identities with the technique

of completing squares in order to get nice factorizations. We will see many examples of this combination of techniques in the sequel, so we only give one famous and very useful example here: Sophie-Germain's classical identity
$$a^4 + 4b^4 = (a^2 - 2ab + 2b^2)(a^2 + 2ab + 2b^2).$$
Of course, you could in principle simply expand the right hand-side and check that the identity holds. However, let us imagine that the problem was: find a factorization of $a^4 + 4b^4$. To solve this problem, we start by completing the square in $a^4 + 4b^4 = (a^2)^2 + (2b^2)^2$ and then complete the factorization using the difference of squares formula $x^2 - y^2 = (x-y)(x+y)$:
$$a^4 + 4b^4 = (a^2)^2 + 2(a^2)(2b^2) + (2b^2)^2 - 4a^2b^2 =$$
$$(a^2 + 2b^2)^2 - (2ab)^2 = (a^2 - 2ab + 2b^2)(a^2 + 2ab + 2b^2).$$

Example 3.1. Simplify the expression
$$\frac{1}{ab} + \frac{1}{a^2 - ab} + \frac{1}{b^2 - ba}.$$

Solution. The key point is that the denominators factor quite nicely $a^2 - ab = a(a-b)$ and $b^2 - ba = b(b-a)$, allowing us to find a common denominator $ab(a-b)$ (you will agree that this is much simpler than the naive product $ab(a^2 - ab)(b^2 - ba)$). Thus
$$\frac{1}{ab} + \frac{1}{a^2 - ab} + \frac{1}{b^2 - ba} = \frac{(a-b) + b - a}{ab(a-b)} = 0.$$

Example 3.2. Simplify
$$\frac{a^2 + 2a - 80}{a^2 - 2a - 120}.$$

Solution. We factor separately the numerator and denominator by completing the square and using the difference of squares formula. We obtain
$$\frac{a^2 + 2a - 80}{a^2 - 2a - 120} = \frac{a^2 + 2a + 1 - 9^2}{a^2 - 2a + 1 - 11^2} = \frac{(a+1)^2 - 9^2}{(a-1)^2 - 11^2} =$$
$$\frac{(a+10)(a-8)}{(a-12)(a+10)} = \frac{a-8}{a-12}.$$

Example 3.3. Solve in real numbers the equation $x^3 - 3x^2 + 3x + 3 = 0$.

Solution. We recognize the successive terms in the expansion
$$(x-1)^3 = x^3 - 3x^2 + 3x - 1.$$
Thus we can write the equation as $(x-1)^3 + 4 = 0$, yielding the unique real solution $x = 1 - \sqrt[3]{4}$.

Factorizations and algebraic identities

Example 3.4. Let x be a real number. Factor the expression $x^4 + x^2 + 1$.

Solution. We start by completing the square, as in the proof of Sophie-Germain's identity

$$x^4 + x^2 + 1 = x^4 + 2x^2 + 1 - x^2 = (x^2 + 1)^2 - x^2.$$

Next we use the formula $a^2 - b^2 = (a - b)(a + b)$ to obtain

$$x^4 + x^2 + 1 = (x^2 - x + 1)(x^2 + x + 1).$$

There is something unsatisfying in the solution of the previous exercise: the most natural way to complete the square would be to see $x^4 + x^2 + 1$ as a quadratic polynomial in $t = x^2$, and then to complete the square as

$$t^2 + t + 1 = \left(t + \frac{1}{2}\right)^2 + \frac{3}{4}.$$

Well, unfortunately we get stuck, since we cannot factor an expression of the form $u^2 + \frac{3}{4}$. Then how can we find the good way of completing the square, so that to be able to further factor the resulting expression? Here is a rather natural approach, which works pretty well in practice: look for a, b such that

$$x^4 + x^2 + 1 = (x^2 + a)^2 - (bx + c)^2$$

for all x. Expanding everything leads to

$$x^4 + x^2 + 1 = x^4 + (2a - b^2)x^2 - 2bcx + a^2 - c^2.$$

If we want this equality to hold for all real numbers, then we must impose

$$2a - b^2 = 1, \quad -2bc = 0, \quad a^2 - c^2 = 1.$$

The second equation gives $b = 0$ or $c = 0$. If $b = 0$, then the first equation yields $a = \frac{1}{2}$ and then we get the contradiction $c^2 + \frac{3}{4} = 0$ (this is exactly the naive approach of completing the square, described above!). Thus $c = 0$, then the last equation gives $a^2 = 1$. Choose for instance $a = 1$. The first equation becomes $b^2 = 1$. Choose $b = 1$. We obtain

$$x^4 + x^2 + 1 = (x^2 + 1)^2 - x^2$$

and we can now apply the difference of squares formula to finish the problem.

Finally, let us give another strategy, which also works very well in practice (see also example 2.9): we remark that the quartic polynomial $x^4 + x^2 + 1$ is symmetric, hence we start by dividing by x^2:

$$\frac{x^4 + x^2 + 1}{x^2} = x^2 + 1 + \frac{1}{x^2}.$$

Next, we let $t = x + \frac{1}{x}$. Then $t^2 = x^2 + 2 + \frac{1}{x^2}$, thus

$$\frac{x^4 + x^2 + 1}{x^2} = t^2 - 1,$$

which factors nicely as $(t-1)(t+1)$. Coming back to the definition of t and multiplying the previous relation by x^2 yields the factorization

$$x^4 + x^2 + 1 = (x^2 - x + 1)(x^2 + x + 1).$$

Let us go on with another example of this phenomenon: completing the squares in the naive way is not always the best decision one could make!

Example 3.5. Factor $(X^4 - 6X^2 + 1)(X^4 - 7X^2 + 1)$.

Solution. The expression is already factored, so either this exercise is a bad joke or there is something more subtle going on! Well, the only subtlety would be that one or both of the expressions $X^4 - 6X^2 + 1$ and $X^4 - 7X^2 + 1$ is itself factorable. It turns out that they both factor: completing the square and using the formula for the difference of squares we obtain

$$X^4 - 6X^2 + 1 = X^4 - 2X^2 + 1 - 4X^2 =$$
$$(X^2 - 1)^2 - (2X)^2 = (X^2 - 2X - 1)(X^2 + 2X - 1)$$

and similarly

$$X^4 - 7X^2 + 1 = X^4 + 2X^2 + 1 - 9X^2 =$$
$$(X^2 + 1)^2 - (3X)^2 = (X^2 + 1 - 3X)(X^2 + 1 + 3X).$$

We finally end up with

$$(X^4 - 6X^2 + 1)(X^4 - 7X^2 + 1)$$
$$= (X^2 - 2X - 1)(X^2 + 2X - 1)(X^2 - 3X + 1)(X^2 + 3X + 1).$$

How to find the previous ways of completing the square? As above, we look for a, b, c so that

$$X^4 - 7X^2 + 1 = (X^2 + a)^2 - (bX + c)^2,$$

we identify coefficients and obtain $c = 0$, $2a - b^2 = -7$ and $a^2 = 1$. In particular $a = 1$ or $a = -1$. If $a = 1$, we get $b^2 = 9$, which is nice, since then we can choose $b = 3$. If $a = -1$, we get $b^2 = 5$, and this is not so nice!

As in the previous example, we can also use the method of symmetric quartic polynomials:

$$\frac{X^4 - 6X^2 + 1}{X^2} = X^2 - 6 + \frac{1}{X^2}.$$

Factorizations and algebraic identities 19

However, we have a problem if we set $t = X + \frac{1}{X}$, since we would end up with $t^2 - 8$, which does not factor nicely (well, it factors as $(t - 2\sqrt{2})(t + 2\sqrt{2})$, but this is not very nice...). Instead we try setting $t = X - \frac{1}{X}$. Then

$$X^2 - 6 + \frac{1}{X^2} = t^2 + 2 - 6 = t^2 - 4 = (t-2)(t+2)$$

and the factorization follows. On the other hand, for $X^4 - 7X^2 + 1$ we don't need this trick (i.e. setting $t = X + \frac{1}{X}$ already works in this case).

Example 3.6. Prove that for all real numbers a, b, c

$$\left(\frac{2a + 2b - c}{3}\right)^2 + \left(\frac{2b + 2c - a}{3}\right)^2 + \left(\frac{2c + 2a - b}{3}\right)^2 = a^2 + b^2 + c^2.$$

Solution. We expand each term in the left hand-side using the formula

$$(x + y + z)^2 = x^2 + y^2 + z^2 + 2xy + 2yz + 2zx.$$

We deduce that the left hand-side equals

$$\frac{4a^2 + 4b^2 + c^2 + 8ab - 4ac - 4bc}{9} + \frac{4b^2 + 4c^2 + a^2 + 8bc - 4ab - 4ac}{9} +$$

$$\frac{4c^2 + 4a^2 + b^2 + 8ac - 4bc - 4ab}{9} = \frac{9a^2 + 9b^2 + 9c^2}{9} = a^2 + b^2 + c^2.$$

One may wonder if we can avoid the previous nasty computation. The answer is, fortunately, positive. Let us denote $S = a + b + c$. Then $2a + 2b - c = 2S - 3c$, so that

$$\left(\frac{2a + 2b - c}{3}\right)^2 = \frac{(2S - 3c)^2}{9} = \frac{4S^2 - 12Sc + 9c^2}{9}.$$

Writing similar equalities for the other two terms and adding them up, we obtain

$$\left(\frac{2a + 2b - c}{3}\right)^2 + \left(\frac{2b + 2c - a}{3}\right)^2 + \left(\frac{2c + 2a - b}{3}\right)^2 =$$

$$\frac{12S^2 - 12S(a + b + c) + 9(a^2 + b^2 + c^2)}{9} = a^2 + b^2 + c^2,$$

since $S(a + b + c) = S^2$.

Example 3.7. Let m and n be positive integers. Prove that

$$(m^3 - 3mn^2)^2 + (3m^2n - n^3)^2$$

is a perfect cube.

Solution. We expand each term using the formula $(a+b)^2 = a^2 + 2ab + b^2$ and rearrange terms, hoping to be able to complete the cube:

$$(m^3 - 3mn^2)^2 + (3m^2n - n^3)^2 = m^6 - 6m^4n^2 + 9m^2n^4 + 9m^4n^2 - 6m^2n^4 + n^6$$

$$= m^6 + 3m^4n^2 + 3m^2n^4 + n^6 = (m^2 + n^2)^3.$$

One could have also observed that

$$(m^3 - 3mn^2)^2 = m^2(m^2 - 3n^2)^2 \quad \text{and} \quad (3m^2n - n^3)^2 = n^2(3m^2 - n^2)^2,$$

hence what really counts is $x = m^2$ and $y = n^2$. This makes the previous computations a little bit simpler and makes it easier to recognize $(m^2 + n^2)^3 = (x+y)^3$ in the last step.

Example 3.8. Solve in integers the equation

$$\frac{x^3 - 3x + 2}{x^2 - 3x + 2} = y.$$

Solution. The good news is that the left hand-side simplifies quite nicely, since both $x^3 - 3x + 2$ and $x^2 - 3x + 2$ vanish at $x = 1$, hence they are both multiples of $x - 1$. Indeed,

$$x^3 - 3x + 2 = x^3 - x - 2(x-1) = x(x+1)(x-1) - 2(x-1) =$$
$$(x^2 + x - 2)(x-1) = (x-1)^2(x+2)$$

and

$$x^2 - 3x + 2 = (x-1)(x-2).$$

Hence the left hand-side equals $\frac{(x+2)(x-1)}{x-2}$. We can simplify this a little bit, since

$$\frac{(x+2)(x-1)}{x-2} = \frac{x^2 + x - 2}{x-2} = \frac{x^2 - 2x + 3(x-2) + 4}{x-2} = x + 3 + \frac{4}{x-2}.$$

Hence we must find all integers x, y such that

$$x + 3 + \frac{4}{x-2} = y.$$

Since $x+3$ and y are integers, so must be $\frac{4}{x-2}$. Thus $x-2$ must be a divisor of 4, that is $x - 2 \in \{-4, -2, -1, 1, 2, 4\}$. Note that $x \neq 1, 2$ since we divided by $x^2 - 3x + 2$. Hence $x \in \{-2, 0, 3, 4, 6\}$. For each such value of x we compute $y = x + 3 + \frac{4}{x-2}$ and we obtain therefore the solutions

$$(x, y) = \{(3, 10), (4, 9), (0, 1), (6, 10), (-2, 0)\}.$$

Factorizations and algebraic identities

Example 3.9. Simplify the expression
$$\frac{x^2 + 4y^2 - z^2 + 4xy}{x^2 - 4y^2 - z^2 + 4yz}.$$

Solution. Let us deal separately with the numerator and the denominator. For the numerator, start by rearranging terms, thus bringing all terms containing x and y together:
$$x^2 + 4y^2 - z^2 + 4xy = x^2 + 4xy + 4y^2 - z^2.$$

It is easy to recognize $x^2 + 4xy + 4y^2$ as $(x+2y)^2$. Hence using the difference of squares formula we obtain
$$x^2 + 4y^2 - z^2 + 4xy = (x+2y)^2 - z^2 = (x+2y-z)(x+2y+z).$$

We can deal similarly with the denominator:
$$x^2 - 4y^2 - z^2 + 4yz = x^2 - (2y-z)^2 = (x-2y+z)(x+2y-z).$$

Hence
$$\frac{x^2 + 4y^2 - z^2 + 4xy}{x^2 - 4y^2 - z^2 + 4yz} = \frac{(x+2y+z)(x+2y-z)}{(x-2y+z)(x+2y-z)} = \frac{x+2y+z}{x-2y+z}.$$

Example 3.10. Solve in real numbers the system of equations
$$\begin{cases} x + y = 2z \\ x^3 + y^3 = 2z^3. \end{cases}$$

Solution. The key ingredient here is the identity
$$x^3 + y^3 = (x+y)(x^2 - xy + y^2).$$

Combining it with the two equations, we obtain
$$2z^3 = 2z(x^2 - xy + y^2).$$

Suppose first that $z = 0$, then the two equations of the system reduce to $x + y = 0$. Thus we obtain the family of solutions $(x, -x, 0)$, with $x \in \mathbf{R}$.

Suppose now that $z \neq 0$. Then the first paragraph yields
$$x^2 - xy + y^2 = z^2.$$

Rewrite
$$x^2 - xy + y^2 = (x+y)^2 - 3xy = 4z^2 - 3xy$$

and compare it with the previous relation. We deduce that $xy = z^2$ and then
$$x^2 - xy + y^2 = z^2 = xy, \quad \text{thus} \quad (x-y)^2 = 0$$

and finally $x = y$. Thus $x = y = z$ and we obtain the second family of solutions (x, x, x), with $x \in \mathbf{R}$.

Example 3.11. Solve in real numbers the system of equations

$$\begin{cases} x + y = xy - 5 \\ y + z = yz - 7 \\ z + x = zx - 11. \end{cases}$$

Solution. The key algebraic identity leading to a short solution is

$$xy - x - y + 1 = (x-1)(y-1).$$

This allows us to write the system as

$$\begin{cases} (x-1)(y-1) = 6 \\ (y-1)(z-1) = 8 \\ (z-1)(x-1) = 12. \end{cases}$$

This strongly suggests changing the variables to

$$a = x - 1, \quad b = y - 1, \quad c = z - 1.$$

Then
$$ab = 6, \quad bc = 8, \quad ca = 12.$$

Taking the product of these equations yields

$$(abc)^2 = 48 \cdot 12 = 4 \cdot 12^2,$$

hence $abc = \pm 24$. If $abc = 24$, then coming back to the equations $ab = 6$, $bc = 8$ and $ca = 12$ we obtain $a = \frac{abc}{bc} = \frac{24}{8} = 3$ and similarly $b = 2$ and $c = 4$. If $abc = -24$, then with the same arguments we obtain $a = -3$, $b = -2$ and $c = -4$. Finally, coming back to the substitutions $a = x - 1$, $b = y - 1$ and $c = z - 1$, we obtain the two solutions $(x, y, z) = (4, 3, 5)$ and $(x, y, z) = (-2, -1, -3)$.

Example 3.12. Let a, b, c, d be real numbers such that

$$a + b + 2ab = 3, \quad b + c + 2bc = 4, \quad c + d + 2cd = -5.$$

Find $d + a + 2da$.

Solution. We can rewrite the equations in a much simpler way by suitably factoring the left hand-side of each equation. The fundamental observation is the factorization

$$1 + 2(x + y + 2xy) = 1 + 2x + 2y + 4xy = (1 + 2x)(1 + 2y).$$

Hence the equations can be rewritten

$$(1+2a)(1+2b) = 7, \quad (1+2b)(1+2c) = 9, \quad (1+2c)(1+2d) = -9.$$

Let us denote

$$x = 2a + 1, \quad y = 2b + 1, \quad z = 2c + 1, \quad t = 2d + 1.$$

Then
$$xy = 7, \quad yz = 9, \quad zt = -9$$

and we need to find
$$d + a + 2da = \frac{xt - 1}{2}.$$

Well, we have $y = \frac{7}{x}$, then $z = \frac{9}{y} = \frac{9x}{7}$ and $\frac{9x}{7} \cdot t = -9$, giving $xt = -7$ and then $d + a + 2da = -4$.

Example 3.13. Solve in positive integers the system of equations

$$\begin{cases} x - y - z = 40 \\ x^2 - y^2 - z^2 = 2012. \end{cases}$$

Solution. Plugging $x = y + z + 40$ into the second equation, we obtain

$$(y + z + 40)^2 - y^2 - z^2 = 2012,$$

and by expanding the first term and simplifying, we obtain the equivalent equation

$$2yz + 80y + 80z = 412, \quad \text{or} \quad yz + 40y + 40z = 206.$$

This can be factored as

$$(y + 40)(z + 40) = 1806.$$

Hence we need to understand the divisors of 1806. We factor it into prime numbers, obtaining
$$1806 = 2 \cdot 3 \cdot 7 \cdot 43 = 42 \cdot 43.$$

Now, since $y + 40$ and $z + 40$ are both greater than 40 (since $y, z > 0$ by assumption) and their product is $42 \cdot 43$, we must have $(y, z) = (2, 3)$ or $(3, 2)$. Since
$$x = y + z + 40 = 5 + 40 = 45,$$
we conclude that the solutions are $(x, y, z) = (45, 2, 3), (45, 3, 2)$.

Example 3.14. Factor $x^4 + 2x^3 + 2x^2 + 2x + 1$.

Solution. The easiest way to deal with the polynomial $p(x) = x^4 + 2x^3 + 2x^2 + 2x + 1$ is to note that it is a symmetric quartic polynomial. Hence we start by dealing with

$$\frac{p(x)}{x^2} = x^2 + 2x + 2 + \frac{2}{x} + \frac{1}{x^2} = x^2 + \frac{1}{x^2} + 2\left(x + \frac{1}{x}\right) + 2.$$

Letting $t = x + \frac{1}{x}$, we have $x^2 + \frac{1}{x^2} = t^2 - 2$ and the previous expression simplifies to

$$t^2 - 2 + 2t + 2 = t^2 + 2t = t(t+2).$$

We deduce that

$$p(x) = x^2 t(t+2) = (x^2+1)(x^2+2x+1) = (x^2+1)(x+1)^2.$$

Here is an alternative approach. The identity

$$u^2 + 2u + 1 = (u+1)^2$$

suggests isolating one x^2 from $2x^2$. We write therefore

$$x^4 + 2x^3 + 2x^2 + 2x + 1 = x^4 + 2x^3 + x^2 + x^2 + 2x + 1.$$

The good news is that $x^4 + 2x^3 + x^2$ is also easily recognizable as $(x^2 + x)^2$ (or, if you prefer, you can start by factoring x^2 and recognizing $(x+1)^2$ as the other factor). Hence

$$x^4 + 2x^3 + 2x^2 + 2x + 1 = (x^2 + x)^2 + (x+1)^2 =$$
$$x^2(x+1)^2 + (x+1)^2 = (x^2+1)(x+1)^2,$$

and this is the desired factorization.

Yet another alternative approach is based on the formula

$$(a+b+c)^2 = a^2 + b^2 + c^2 + 2ab + 2bc + 2ca,$$

the hint being that many of the coefficients of the given expression are equal to 2. Contemplating the previous identity and the expression $E = x^4 + 2x^3 + 2x^2 + 2x + 1$, we see that $(x^2 + x + 1)^2$ is closely related to it. Indeed,

$$(x^2 + x + 1)^2 = x^4 + 2x^3 + 3x^2 + 2x + 1 = E + x^2.$$

Thus

$$E = (x^2+x+1)^2 - x^2 = (x^2+x+1-x)(x^2+x+1+x) = (x^2+1)(x+1)^2.$$

Example 3.15. a) Factor the expression $x^5 + x + 1$.
  b) Find the prime factorization of 100011.

Solution. a) This is fairly tricky: we add and subtract x^2 to get
$$x^5 + x + 1 = x^5 - x^2 + x^2 + x + 1 = x^2(x^3 - 1) + x^2 + x + 1.$$
The advantage is that $x^3 - 1$ is a multiple of $x^2 + x + 1$, hence
$$x^2(x^3-1)+x^2+x+1 = (x^2+x+1)(x^2(x-1)+1) = (x^2+x+1)(x^3-x^2+1).$$
We conclude that
$$x^5 + x + 1 = (x^2 + x + 1)(x^3 - x^2 + 1).$$

b) We apply part a) to $x = 10$ and obtain
$$100011 = 10^5 + 10 + 1 = (1000 - 100 + 1)(100 + 10 + 1) = 901 \cdot 111 = 3 \cdot 17 \cdot 37 \cdot 53.$$

Example 3.16. Solve in real numbers the equation
$$\frac{1}{x+1} + \frac{1}{x^2 - x} + \frac{1}{x^3 - x} = 1.$$

Solution. We start by finding a good common denominator. In order to do that, we factor each denominator: $x^2 - x = x(x - 1)$ and
$$x^3 - x = x(x^2 - 1) = x(x - 1)(x + 1).$$
Hence $x(x-1)(x+1) = x^3 - x$ is a common denominator. Multiplying the numerator and denominator of the first (resp. second) fraction by $x(x-1)$ (resp. $x+1$) we obtain the equivalent equation
$$\frac{x(x-1) + x + 1 + 1}{x^3 - x} = 1$$
and then $x^3 - x^2 - x - 2 = 0$.

Now we are quite stuck! Well, we look for integer solutions. These are among the divisors of 2 and their negatives, so trial and error gives the solution $x = 2$. Next, we rewrite the cubic equation in order to factor $x - 2$:
$$x^3 - 2x^2 + x^2 - 2x + x - 2 = 0 \quad \text{or} \quad (x - 2)(x^2 + x + 1) = 0.$$

The equation $x^2 + x + 1 = 0$ has no real solutions, since the discriminant is negative. Hence the only solution of the equation is $x = 2$ (we easily check that $x + 1 \neq 0$, $x^2 \neq x$ and $x^3 \neq x$ so $x = 2$ really is a solution).

Example 3.17. Let a, b, c be real numbers such that $ab + bc + ca = 1$. Prove that
$$\frac{a}{a^2 + 1} + \frac{b}{b^2 + 1} + \frac{c}{c^2 + 1} = \frac{2}{(a+b)(b+c)(c+a)}.$$

Solution. The key point is replacing 1 in each denominator by $ab+bc+ca$. Why is this useful? Well, because of the factorization

$$a^2 + (ab+bc+ca) = a^2 + ab + bc + ca = a(a+b) + c(a+b) = (a+b)(a+c)$$

and similar ones obtained by permutations of the variables. Hence

$$\frac{a}{a^2+1} + \frac{b}{b^2+1} + \frac{c}{c^2+1} = \frac{a}{(a+b)(a+c)} + \frac{b}{(b+a)(b+c)} + \frac{c}{(c+a)(c+b)} =$$

$$\frac{a(b+c) + b(c+a) + c(a+b)}{(c+a)(c+b)(a+b)} = \frac{2}{(a+b)(b+c)(c+a)},$$

the last equality being again a consequence of the hypothesis $ab+bc+ca = 1$.

Example 3.18. If a, b, c, d are real numbers such that $a^2 + b^2 + c^2 + d^2 \leq 1$, find the maximal value of

$$(a+b)^4 + (a+c)^4 + (a+d)^4 + (b+c)^4 + (b+d)^4 + (c+d)^4.$$

Solution. This problem is also fairly challenging. We will try to add some extra terms to the expression to make it an even polynomial in a, b, c, d, as then it could be expressed only in terms of a^2, b^2, c^2, d^2. The key point is that

$$(a+b)^4 + (a-b)^4 = a^4 + 4a^3b + 6a^2b^2 + 4ab^3 + b^4 + a^4 -$$

$$4a^3b + 6a^2b^2 - 4ab^3 + b^4 = 2(a^4 + b^4 + 6a^2b^2)$$

is an even polynomial in a, b. The previous identity also simplifies drastically the following expression:

$$(a+b)^4 + (a+c)^4 + (a+d)^4 + (b+c)^4 + (b+d)^4 + (c+d)^4 + (a-b)^4 + (a-c)^4$$

$$+ (a-d)^4 + (b-c)^4 + (b-d)^4 + (b-d)^4 + (c-d)^4 = 6(a^4 + b^4 + c^4 + d^4)$$

$$+ 12(a^2b^2 + b^2c^2 + c^2d^2 + d^2a^2 + a^2c^2 + b^2d^2).$$

We recognize the identity

$$(x+y+z+t)^2 = x^2 + y^2 + z^2 + t^2 + 2(xy + yz + zt + tx + xz + yt)$$

in the last expression, which is therefore equal to $6(a^2+b^2+c^2+d^2)^2 \leq 6$. We conclude that

$$(a+b)^4 + (a+c)^4 + (a+d)^4 + (b+c)^4 + (b+d)^4 + (c+d)^4 \leq 6,$$

with equality when $a = b = c = d$ and $a^2 + b^2 + c^2 + d^2 = 1$, that is $a = b = c = d = \pm\frac{1}{2}$. Thus the answer is 6.

4 Factoring expressions involving $a - b$, $b - c$, $c - a$

The fundamental property of the numbers $a - b$, $b - c$ and $c - a$ is that they add up to zero:
$$(a - b) + (b - c) + (c - a) = a - b + b - c + c - a = 0.$$

The converse is true: if x, y, z are real numbers adding up to zero, then we can find real numbers a, b, c such that
$$x = a - b, \quad y = b - c, \quad z = c - a.$$

Indeed, we can take $a = 0, b = -x$ and $c = z$. Note that a, b, c are not unique, since we may add up some common number to each of them, obtaining a triple which satisfies the same equalities as (a, b, c).

One encounters quite often symmetric polynomial expressions of the form
$$(a - b)P(a, b, c) + (b - c)Q(a, b, c) + (c - a)R(a, b, c),$$

where P, Q, R are polynomials (or even rational fractions), and very often such expressions have nice factorizations. A method that works quite nicely to factor these expressions is to express $c - a$ as $-((a - b) + (b - c))$ and then to separate terms according to $a - b$ and $b - c$. The following examples should convince the reader that this is a rather versatile tool.

Example 4.1. Factor $a^2(b - c) + b^2(c - a) + c^2(a - b)$ and find a necessary and sufficient condition for a, b, c to satisfy
$$a^2 b + b^2 c + c^2 a = ab^2 + bc^2 + ca^2.$$

Solution. We replace $c - a$ with $-((a - b) + (b - c))$ and get
$$a^2(b-c) + b^2(c-a) + c^2(a-b) = a^2(b-c) - b^2(b-c) - b^2(a-b) + c^2(a-b) =$$
$$(a^2 - b^2)(b - c) - (b^2 - c^2)(a - b).$$

The good news is that we are in pretty good shape now, due to the presence of $a^2 - b^2$ and $b^2 - c^2$, which factor out very nicely. Hence we obtain
$$a^2(b-c) + b^2(c-a) + c^2(a-b) = (a-b)(b-c)(a+b) - (b-c)(a-b)(b+c) =$$
$$(a-b)(b-c)(a+b-b-c) = (a-b)(b-c)(a-c),$$

that is
$$a^2(b - c) + b^2(c - a) + c^2(a - b) = -(a - b)(b - c)(c - a).$$

In particular, we have
$$a^2 b + b^2 c + c^2 a = ab^2 + bc^2 + ca^2$$

if and only if two of the numbers a, b, c are equal.

Example 4.2. Factor $a^4(b-c)+b^4(c-a)+c^4(a-b)$ and find all real numbers a,b,c for which
$$a^4(b-c)+b^4(c-a)+c^4(a-b)=0.$$

Solution. Replacing $a-b$ by $a-c-(b-c)$, we obtain
$$a^4(b-c)+b^4(c-a)+c^4(a-b)=(b-c)(a^4-c^4)+(c-a)(b^4-c^4)=$$
$$(b-c)(a-c)(a^3+a^2c+ac^2+c^3)+(c-a)(b-c)(b^3+b^2c+bc^2+c^3)=$$
$$(b-c)(c-a)(b^3+b^2c+bc^2+c^3-a^3-a^2c-ac^2-c^3).$$

Note that $b^3+b^2c+bc^2+c^3-a^3-a^2c-ac^2-c^3$ vanishes when $a=b$, hence $a-b$ must divide it. Indeed,
$$b^3+b^2c+bc^2+c^3-a^3-a^2c-ac^2-c^3=b^3-a^3+(b^2-a^2)c+(b-a)c^2=$$
$$(b-a)(b^2+ab+a^2+(b+a)c+c^2)=(b-a)(a^2+b^2+c^2+ab+bc+ca).$$

It follows that
$$a^4(b-c)+b^4(c-a)+c^4(a-b)=-(a-b)(b-c)(c-a)(a^2+b^2+c^2+ab+bc+ca).$$

To answer the second question, we use what we have already established and write the condition as
$$(a-b)(b-c)(c-a)(a^2+b^2+c^2+ab+bc+ca)=0.$$

This happens if and only if two of the numbers a,b,c are equal or
$$a^2+b^2+c^2+ab+bc+ca=0.$$

Multiplying by 2, the previous relation becomes
$$(a+b)^2+(b+c)^2+(c+a)^2=0$$
and forces
$$a+b=b+c=c+a=0 \quad \text{hence} \quad a=b=c=0.$$

We conclude that
$$a^4(b-c)+b^4(c-a)+c^4(a-b)=0$$
if and only if two of the numbers a,b,c are equal.

Example 4.3. Factor
$$(a-b)(a^2+b^2-c^2)c^2+(b-c)(b^2+c^2-a^2)a^2+(c-a)(c^2+a^2-b^2)b^2.$$

Solution. Let us denote $s = a^2 + b^2 + c^2$. Then the given expression equals

$$(a-b)(s-2c^2)c^2 + (b-c)(s-2a^2)a^2 + (c-a)(s-2b^2)b^2 =$$
$$s\left[a^2(b-c) + b^2(c-a) + c^2(a-b)\right] - 2\left[a^4(b-c) + b^4(c-a) + c^4(a-b)\right].$$

Using the previous two exercises, we obtain

$$(a-b)(a^2+b^2-c^2)c^2 + (b-c)(b^2+c^2-a^2)a^2 + (c-a)(c^2+a^2-b^2)b^2 =$$
$$-(a-b)(b-c)(c-a)(a^2+b^2+c^2) + 2(a-b)(b-c)(c-a)(a^2+b^2+c^2+ab+bc+ca) =$$
$$(a-b)(b-c)(c-a)(a^2+b^2+c^2+2ab+2bc+2ca)$$
$$= (a-b)(b-c)(c-a)(a+b+c)^2.$$

For the reader who didn't notice the trick of denoting $s = a^2 + b^2 + c^2$ and thus reducing the problem to the previous two exercises, we give an alternative and direct way of handling the problem. We start by replacing $c-a$ by $-(a-b)-(b-c)$. The expression becomes

$$(a-b)[(a^2+b^2-c^2)c^2 - (c^2+a^2-b^2)b^2] + (b-c)[(b^2+c^2-a^2)a^2 - (c^2+a^2-b^2)b^2].$$

Next, we deal separately with the large terms inside the brackets. We have

$$(a^2+b^2-c^2)c^2 - (c^2+a^2-b^2)b^2 = a^2c^2 + b^2c^2 - c^4 - b^2c^2 - a^2b^2 + b^4 =$$
$$a^2c^2 - c^4 - a^2b^2 + b^4 = a^2(c^2-b^2) - (c^2-b^2)(c^2+b^2) = (c^2-b^2)(a^2-b^2-c^2).$$

Similarly, we obtain

$$(b^2+c^2-a^2)a^2 - (c^2+a^2-b^2)b^2 = (a^2-b^2)(c^2-a^2-b^2).$$

Thus

$$(a-b)(a^2+b^2-c^2)c^2 + (b-c)(b^2+c^2-a^2)a^2 + (c-a)(c^2+a^2-b^2)b^2 =$$
$$(a-b)(c^2-b^2)(a^2-b^2-c^2) + (b-c)(a^2-b^2)(c^2-a^2-b^2) =$$
$$(a-b)(b-c)(-(b+c)(a^2-b^2-c^2) + (a+b)(c^2-a^2-b^2)).$$

Now let us focus on the term

$$-(b+c)(a^2-b^2-c^2) + (a+b)(c^2-a^2-b^2) =$$
$$(b+c)(a^2+b^2+c^2-2a^2) - (a+b)(a^2+b^2+c^2-2c^2) =$$
$$(a^2+b^2+c^2)(c-a) - 2a^2(b+c) + 2c^2(a+b).$$

Finally,
$$2c^2(a+b) - 2a^2(b+c) = 2c^2a + 2c^2b - 2a^2b - 2a^2c =$$
$$2ac(c-a) + 2b(c-a)(c+a) = 2(c-a)(ab+bc+ca).$$

Replacing this in the previous expressions yields
$$(a-b)(a^2+b^2-c^2)c^2 + (b-c)(b^2+c^2-a^2)a^2 + (c-a)(c^2+a^2-b^2)b^2 =$$
$$(a-b)(b-c)(c-a)(a^2+b^2+c^2+2ab+2bc+2ca) = (a-b)(b-c)(c-a)(a+b+c)^2$$

and we are finally done!

Example 4.4. Prove that for all real numbers a, b, c such that $(a+b)(b+c)(c+a) \neq 0$ we have

$$\frac{a-b}{a+b} + \frac{b-c}{b+c} + \frac{c-a}{c+a} = -\frac{a-b}{a+b} \cdot \frac{b-c}{b+c} \cdot \frac{c-a}{c+a}.$$

Solution. We replace $c - a$ with $-((a-b) + (b-c))$ to obtain

$$\frac{a-b}{a+b} + \frac{b-c}{b+c} + \frac{c-a}{c+a} = (a-b)\left(\frac{1}{a+b} - \frac{1}{c+a}\right) + (b-c)\left(\frac{1}{b+c} - \frac{1}{c+a}\right) =$$

$$(a-b)\frac{c-b}{(a+b)(a+c)} + (b-c)\frac{a-b}{(b+c)(c+a)} = \frac{(a-b)(b-c)}{a+c}\left(\frac{1}{b+c} - \frac{1}{a+b}\right) =$$

$$\frac{(a-b)(b-c)}{a+c} \cdot \frac{a-c}{(a+b)(b+c)} = -\frac{a-b}{a+b} \cdot \frac{b-c}{b+c} \cdot \frac{c-a}{c+a}.$$

The result follows.

Example 4.5. Prove that for all pairwise distinct real numbers a, b, c we have

$$\frac{a+b}{a-b} \cdot \frac{a+c}{a-c} + \frac{b+c}{b-c} \cdot \frac{b+a}{b-a} + \frac{c+a}{c-a} \cdot \frac{c+b}{c-b} = 1.$$

Solution. The most direct way to solve this problem is to divide by the right hand-side the identity in the previous exercise (actually this will deal with the case when $(a+b)(b+c)(c+a) \neq 0$, but if $a+b = 0$ the result is rather easy to establish). Let us not give a proof independent of the previous exercise. We write

$$\frac{a+b}{a-b} \cdot \frac{a+c}{a-c} + \frac{b+c}{b-c} \cdot \frac{b+a}{b-a} + \frac{c+a}{c-a} \cdot \frac{c+b}{c-b} =$$

$$-\frac{(a+b)(a+c)(b-c) + (b+c)(b+a)(c-a) + (c+a)(c+b)(a-b)}{(a-b)(b-c)(c-a)}.$$

Next, denoting $s = ab + bc + ca$, we deal with the numerator as follows:

$$(a+b)(a+c)(b-c) + (b+c)(b+a)(c-a) + (c+a)(c+b)(a-b)$$
$$= (a^2 + s)(b-c) + (b^2 + s)(c-a) + (c^2 + s)(a-b)$$
$$= s(b-c+c-a+a-b) + a^2(b-c) + b^2(b-c) + c^2(a-b)$$
$$= -(a-b)(b-c)(c-a),$$

the last identity being a consequence of exercise 1. We conclude that

$$\frac{a+b}{a-b} \cdot \frac{a+c}{a-c} + \frac{b+c}{b-c} \cdot \frac{b+a}{b-a} + \frac{c+a}{c-a} \cdot \frac{c+b}{c-b} =$$

$$\frac{(a-b)(b-c)(c-a)}{(a-b)(b-c)(c-a)} = 1.$$

Example 4.6. Prove that for all pairwise distinct real numbers a, b, c

$$\frac{bc}{(a-b)(a-c)} + \frac{ca}{(b-c)(b-a)} + \frac{ab}{(c-a)(c-b)} = 1.$$

Solution. The expression on the left hand-side is equal to

$$-\frac{bc(b-c) + ca(c-a) + ab(a-b)}{(a-b)(b-c)(c-a)}.$$

But

$$bc(b-c) + ca(c-a) + ab(a-b) = b^2c - bc^2 + c^2a - ca^2 + a^2b - ab^2 =$$
$$a^2(b-c) + b^2(c-a) + c^2(a-b) = -(a-b)(b-c)(c-a),$$

where the last relation follows from exercise 1. We conclude that

$$\frac{bc}{(a-b)(a-c)} + \frac{ca}{(b-c)(b-a)} + \frac{ab}{(c-a)(c-b)}$$
$$= \frac{(a-b)(b-c)(c-a)}{(a-b)(b-c)(c-a)} = 1.$$

Example 4.7. Prove that if a, b, c are pairwise distinct real numbers, then

$$\frac{a^2}{(b-c)^2} + \frac{b^2}{(c-a)^2} + \frac{c^2}{(a-b)^2} \geq 2.$$

Solution. Using the previous exercise, we can write

$$\frac{a^2}{(b-c)^2} + \frac{b^2}{(c-a)^2} + \frac{c^2}{(a-b)^2} = \left(\frac{a}{b-c} + \frac{b}{c-a} + \frac{c}{a-b}\right)^2 +$$

$$2\left(\frac{bc}{(a-b)(a-c)} + \frac{ca}{(b-c)(b-a)} + \frac{ab}{(c-a)(c-b)}\right) =$$
$$\left(\frac{a}{b-c} + \frac{b}{c-a} + \frac{c}{a-b}\right)^2 + 2 \geq 2$$

and we are done.

Example 4.8. Prove that if a, b, c are pairwise distinct real numbers, then
$$\frac{a-b}{1+ab} + \frac{b-c}{1+bc} + \frac{c-a}{1+ca} \neq 0.$$

Solution. Again, replacing $c - a$ by $-((a-b) + (b-c))$ yields a nice factorization of the left hand-side:

$$\frac{a-b}{1+ab} + \frac{b-c}{1+bc} + \frac{c-a}{1+ca}$$
$$= (a-b)\left(\frac{1}{1+ab} - \frac{1}{1+ca}\right) + (b-c)\left(\frac{1}{1+bc} - \frac{1}{1+ca}\right)$$
$$= (a-b)\frac{a(c-b)}{(1+ab)(1+ac)} + (b-c)\frac{c(a-b)}{(1+bc)(1+ca)}$$
$$= \frac{(a-b)(b-c)}{(1+ab)(1+bc)(1+ca)}(c(1+ab) - a(1+bc))$$
$$= \frac{(a-b)(b-c)(c-a)}{(1+ab)(1+bc)(1+ca)}.$$

The result is now clear, since by hypothesis $(a-b)(b-c)(c-a) \neq 0$.

Example 4.9. Prove that for all real numbers a, b, c we have
$$(a-b)^5 + (b-c)^5 + (c-a)^5 = 5(a-b)(b-c)(c-a)(a^2+b^2+c^2-ab-bc-ca).$$

Solution. Let us set
$$x = a - b, \quad y = b - c, \quad z = c - a.$$

Then $x + y + z = 0$ and
$$a^2 + b^2 + c^2 - ab - bc - ca = \frac{(a-b)^2 + (b-c)^2 + (c-a)^2}{2} = \frac{x^2 + y^2 + z^2}{2}.$$

Hence we need to prove that
$$x^5 + y^5 + z^5 = \frac{5}{2}xyz(x^2 + y^2 + z^2).$$

Well, we replace z by $-x - y$ and expand:

$$x^5 + y^5 + z^5 = x^5 + y^5 - (x+y)^5 = (x+y)(x^4 - x^3y + x^2y^2 - xy^3 + y^4 - (x+y)^4) =$$
$$(x+y)(-x^3y + x^2y^2 - xy^3 - 4x^3y - 6x^2y^2 - 4xy^3) = -5(x+y)(x^3y + xy^3 + x^2y^2) =$$
$$5zxy(x^2 + y^2 + xy).$$

On the other hand,

$$\frac{5}{2}xyz(x^2 + y^2 + z^2) = \frac{5}{2}xyz(x^2 + y^2 + (x+y)^2) = 5xyz(x^2 + xy + y^2).$$

Comparing the two expressions yields the desired result.

Here is an alternative approach, which exploits the symmetry of the problem: let $S = xy + yz + zx$ and $P = xyz$. Then

$$(t - x)(t - y)(t - z) = t^3 + St - P$$

for all real numbers t. In particular, x, y, z are all roots of the equation $t^3 + St - P = 0$, hence also of the equation $t^5 + St^3 - Pt^2 = 0$. Combining these two equations, we deduce that x, y, z are roots of the equation $t^5 + S(P - St) - Pt^2 = 0$. Adding these relations yields

$$x^5 + y^5 + z^5 + S(3P - S(x + y + z)) - P(x^2 + y^2 + z^2) = 0.$$

Since $x + y + z = 0$, this becomes

$$x^5 + y^5 + z^5 + 3SP - P(x^2 + y^2 + z^2) = 0$$

or

$$x^5 + y^5 + z^5 = P(x^2 + y^2 + z^2 - 3S).$$

Hence it suffices to prove that

$$x^2 + y^2 + z^2 - 3(xy + yz + zx) = \frac{5}{2}(x^2 + y^2 + z^2).$$

This is equivalent to

$$x^2 + y^2 + z^2 + 2(xy + yz + zx) = 0,$$

which is simply the relation $(x + y + z)^2 = 0$.

Example 4.10. Positive real numbers a, b, c satisfy

$$\frac{a(b - c)}{b + c} + \frac{b(c - a)}{c + a} + \frac{c(a - b)}{a + b} = 0.$$

Prove that $(a - b)(b - c)(c - a) = 0$.

Solution. Replace $c-a$ by $c-b+b-a$ and rewrite the equality as

$$\left(\frac{a}{b+c}-\frac{b}{c+a}\right)(b-c)+\left(\frac{c}{a+b}-\frac{b}{c+a}\right)(a-b)=0.$$

Next, observe that

$$\frac{a}{b+c}-\frac{b}{c+a}=\frac{ac+a^2-b^2-bc}{(b+c)(c+a)}=\frac{(a-b)(a+b+c)}{(b+c)(c+a)}.$$

Doing the same thing with $\frac{c}{a+b}-\frac{b}{c+a}$, we obtain the equivalent relation

$$(a+b+c)\cdot\frac{(a-b)(b-c)}{(b+c)(c+a)}-(a+b+c)\cdot\frac{(a-b)(b-c)}{(a+b)(a+c)}=0.$$

Suppose that $(a-b)(b-c)(c-a)\neq 0$. Dividing the previous equality by $(a+b+c)(a-b)(a-c)$ (note that $a+b+c\neq 0$, since a,b,c are positive) yields

$$\frac{1}{(b+c)(c+a)}=\frac{1}{(a+b)(a+c)},$$

which is easily seen to be equivalent to $a=c$. The result follows.

Example 4.11. Prove that for all pairwise distinct real numbers a,b,c

$$a^2\frac{(a+b)(a+c)}{(a-b)(a-c)}+b^2\frac{(b+c)(b+a)}{(b-c)(b-a)}+c^2\frac{(c+a)(c+b)}{(c-a)(c-b)}=(a+b+c)^2.$$

Solution. The left hand-side equals

$$-\frac{a^2(a+b)(a+c)(b-c)+b^2(b+c)(b+a)(c-a)+c^2(c+a)(c+b)(a-b)}{(a-b)(b-c)(c-a)}.$$

On the other hand, denoting $s=ab+bc+ca$, we can write

$$a^2(a+b)(a+c)(b-c)+b^2(b+c)(b+a)(c-a)+c^2(c+a)(c+b)(a-b)=$$

$$a^2(a^2+s)(b-c)+b^2(b^2+s)(c-a)+c^2(c^2+s)(a-b)=$$

$$a^4(b-c)+b^4(c-a)+c^4(a-b)+s\left[a^2(b-c)+b^2(c-a)+c^2(a-b)\right].$$

The result follows then easily by combining exercises 1 and 2.

Example 4.12. Prove that for all positive real numbers a,b,c

$$\frac{a^2(b+c)}{b^2+c^2}+\frac{b^2(c+a)}{c^2+a^2}+\frac{c^2(a+b)}{a^2+b^2}\geq a+b+c.$$

Solution. This problem is fairly difficult. We write the difference between the left hand-side and the right hand-side as

$$\frac{a^2(b+c)}{b^2+c^2} - a + \frac{b^2(c+a)}{c^2+a^2} - b + \frac{c^2(a+b)}{a^2+b^2} - c$$

$$= a \cdot \frac{ab+ac-b^2-c^2}{b^2+c^2} + b \cdot \frac{bc+ab-c^2-a^2}{c^2+a^2} + c \cdot \frac{ca+bc-a^2-b^2}{a^2+b^2}$$

$$= \frac{ab(a-b)+ac(a-c)}{b^2+c^2} + \frac{bc(b-c)+ab(b-a)}{c^2+a^2} + \frac{ca(c-a)+bc(c-b)}{a^2+b^2}$$

$$= ab(a-b)\left(\frac{1}{b^2+c^2} - \frac{1}{c^2+a^2}\right) + bc(b-c)\left(\frac{1}{c^2+a^2} - \frac{1}{a^2+b^2}\right)$$

$$+ ca(c-a)\left(\frac{1}{a^2+b^2} - \frac{1}{b^2+c^2}\right).$$

We claim that each of the terms in the previous sum is nonnegative, which is enough to conclude. Indeed, we have

$$ab(a-b)\left(\frac{1}{b^2+c^2} - \frac{1}{c^2+a^2}\right) = \frac{ab(a-b)(a^2-b^2)}{(c^2+a^2)(c^2+b^2)} =$$

$$\frac{ab(a-b)^2(a+b)}{(c^2+a^2)(c^2+b^2)} \geq 0.$$

The argument being the same for the other two terms, the result follows.

5 Factoring $a^3 + b^3 + c^3 - 3abc$

A very useful algebraic identity is

$$a^3 + b^3 + c^3 - 3abc = (a+b+c)(a^2 + b^2 + c^2 - ab - bc - ca).$$

The proof of this identity is fairly easy: simply expand the right hand-side and simplify the resulting expression. Let us remark that

$$a^2 + b^2 + c^2 - ab - bc - ca = \frac{(a-b)^2 + (b-c)^2 + (c-a)^2}{2}$$

is nonnegative for all real numbers a, b, c, and that it equals 0 if and only if $a = b = c$. Hence

$$a^3 + b^3 + c^3 = 3abc$$

holds if and only if $a = b = c$ or $a + b + c = 0$. Let us see a few examples of applications of this algebraic identity.

Example 5.1. Prove that for all real numbers a, b, c we have

$$(a-b)^3 + (b-c)^3 + (c-a)^3 = 3(a-b)(b-c)(c-a).$$

Solution. It is rather clear that what really plays a role here are the numbers $a - b, b - c$ and $c - a$, rather than a, b, c. Thus we start by making the substitution $x = a - b, y = b - c$ and $z = c - a$. We need to prove that

$$x^3 + y^3 + z^3 = 3xyz.$$

Now we use the identity

$$x^3 + y^3 + z^3 - 3xyz = (x+y+z)(x^2 + y^2 + z^2 - xy - yz - xz).$$

Well, the good news is that

$$x + y + z = a - b + b - c + c - a = 0,$$

hence we are done.

Example 5.2. Let a, b, c be real numbers such that

$$(a-b)^2 + (b-c)^2 + (c-a)^2 = 6.$$

Prove that

$$a^3 + b^3 + c^3 = 3(a + b + c + abc).$$

Solution. Write the desired equality as
$$a^3 + b^3 + c^3 - 3abc = 3(a+b+c).$$
The left hand-side factors as
$$(a+b+c)(a^2+b^2+c^2-ab-bc-ca) =$$
$$(a+b+c)\frac{(a-b)^2+(b-c)^2+(c-a)^2}{2} = 3(a+b+c),$$
the last equality being a consequence of the hypothesis.

Example 5.3. Let a, b, c be real numbers such that $a+b+c=1$. Prove that
$$a^3 + b^3 + c^3 - 1 = 3(abc - ab - bc - ca).$$

Solution. Let us rearrange a little bit the equality, so as to make $a^3 + b^3 + c^3 - 3abc$ appear:
$$a^3 + b^3 + c^3 - 3abc = 1 - 3(ab + bc + ca).$$
The left hand-side factors as
$$(a+b+c)(a^2+b^2+c^2-ab-bc-ca) = a^2+b^2+c^2-ab-bc-ca,$$
the last equality taking into account the hypothesis that $a+b+c=1$. Thus it suffices to prove that
$$a^2 + b^2 + c^2 - ab - bc - ca = 1 - 3(ab + bc + ca),$$
which can be rewritten as
$$a^2 + b^2 + c^2 + 2ab + 2bc + 2ca = 1.$$
But we easily recognize that the left hand-side is simply $(a+b+c)^2 = 1$.

Example 5.4. Let a, b, c be real numbers such that
$$\left(-\frac{a}{2} + \frac{b}{3} + \frac{c}{6}\right)^3 + \left(\frac{a}{3} + \frac{b}{6} - \frac{c}{2}\right)^3 + \left(\frac{a}{6} - \frac{b}{2} + \frac{c}{3}\right)^3 = \frac{1}{8}.$$
Prove that
$$(a - 3b + 2c)(2a + b - 3c)(-3a + 2b + c) = 9.$$

Solution. Both the hypothesis and the conclusion look fairly complicated, so let us start by writing the hypothesis in a more convenient way. Clearing denominators, we obtain the equivalent equality

$$\left(\frac{-3a+2b+c}{6}\right)^3 + \left(\frac{2a+b-3c}{6}\right)^3 + \left(\frac{a-3b+2c}{6}\right)^3 = \frac{1}{8}$$

or

$$(-3a+2b+c)^3 + (2a+b-3c)^3 + (a-3b+2c)^3 = 27.$$

The good news is that everything depends only on the variables

$$x = -3a+2b+c, \quad y = 2a+b-3c, \quad z = -3a+2b+c.$$

Namely, the hypothesis can be written as

$$x^3 + y^3 + z^3 = 27$$

and the conclusion as $xyz = 9$. Of course, it is not true that if x, y, z are arbitrary real numbers with $x^3 + y^3 + z^3 = 27$, then $xyz = 9$. So there is probably a further relation satisfied by x, y, z. This relation is fairly apparent: adding up the previous relations yields

$$x + y + z = -3a + 2b + c + 2a + b - 3c - 3a + 2b + c = 0.$$

We are now in good shape, since the identity

$$x^3 + y^3 + z^3 - 3xyz = (x+y+z)(x^2+y^2+z^2 - xy - xz - yx) = 0$$

combined with the hypothesis and with what we have already proved yields $3xyz = 27$ and then $xyz = 9$.

Example 5.5. Find all real numbers a, b, c such that

$$\sqrt[3]{a-b} + \sqrt[3]{b-c} + \sqrt[3]{c-a} = 0.$$

Solution. Let $x = \sqrt[3]{a-b}$, $y = \sqrt[3]{b-c}$ and $z = \sqrt[3]{c-a}$. Then $x+y+z = 0$ by hypothesis, and clearly

$$x^3 + y^3 + z^3 = a - b + b - c + c - a = 0.$$

Hence the identity

$$x^3 + y^3 + z^3 - 3xyz = (x+y+z)(x^2+y^2+z^2-xy-yz-zx)$$

yields $3xyz = 0$. Suppose without loss of generality that $x = 0$, then $a = b$ and the desired relation is trivially satisfied. Hence the answer is: all triples (a, b, c) for which $a = b$ or $b = c$ or $c = a$.

Example 5.6. Prove that if a, b, c, d are real numbers satisfying $a+b+c+d = 0$, then
$$a^3 + b^3 + c^3 + d^3 = 3(abc + bcd + cda + dab).$$

Solution. Using the hypothesis, we can write
$$a^3 + b^3 + c^3 - 3abc = (a+b+c)(a^2 + b^2 + c^2 - ab - bc - ca) =$$
$$-d(a^2 + b^2 + c^2) + dab + dbc + dca.$$

We write down three similar equalities, obtained by permuting cyclically a, b, c, d, and we add up the resulting relations, ending up with
$$3(a^3 + b^3 + c^3 + d^3) - 3(abc + bcd + cda + dab) =$$
$$3(abc+bcd+cda+dab)-a(b^2+c^2+d^2)-b(a^2+c^2+d^2)-c(a^2+b^2+d^2)-d(a^2+b^2+c^2).$$

On the other hand, we have
$$a(b^2 + c^2 + d^2) + b(a^2 + c^2 + d^2) + c(a^2 + b^2 + d^2) + d(a^2 + b^2 + c^2) =$$
$$a^2(b+c+d) + b^2(c+d+a) + c^2(a+b+d) + d^2(a+b+c)$$
$$= -a^3 - b^3 - c^3 - d^3,$$

where the first equality follows by rearranging terms and the second equality is a consequence of the hypothesis $a + b + c + d = 0$. Thus we end up with
$$3(a^3 + b^3 + c^3 + d^3) - 3(abc + bcd + cda + dab)$$
$$= 3(abc + bcd + cda + dab) + (a^3 + b^3 + c^3 + d^3),$$

which after simplifications is exactly the desired equality.

Example 5.7. Let a, b, c be nonzero real numbers, not all equal, such that
$$\frac{1}{a} + \frac{1}{b} + \frac{1}{c} = 1 \quad \text{and} \quad a^3 + b^3 + c^3 = 3(a^2 + b^2 + c^2).$$
Prove that $a + b + c = 3$.

Solution. The first relation can be written as
$$3abc = 3(ab + bc + ca),$$
so
$$a^3 + b^3 + c^3 - 3abc = 3(a^2 + b^2 + c^2 - ab - bc - ca).$$

Noting that
$$a^2 + b^2 + c^2 - ab - bc - ca = \frac{(a-b)^2 + (b-c)^2 + (c-a)^2}{2} \neq 0,$$
the result follows.

6 AM-GM and Hölder's inequality

The easiest, but also probably the most important case of the AM-GM inequality is
$$a + b \geq 2\sqrt{ab},$$
which holds for all nonnegative real numbers a, b. This inequality is equivalent to $(\sqrt{a} - \sqrt{b})^2 \geq 0$, thus clearly true. We also see that we have equality if and only if $\sqrt{a} = \sqrt{b}$, that is $a = b$. We can also write it as
$$\frac{a^2 + b^2}{2} \geq ab,$$
the advantage of this last inequality being that it holds for all real numbers a, b (since it is equivalent to $(a - b)^2 \geq 0$).

Let us take now four nonnegative numbers a, b, c, d and apply the previous inequality twice. We obtain
$$a + b + c + d \geq 2\sqrt{ab} + 2\sqrt{cd} = 2(\sqrt{ab} + \sqrt{cd})$$
$$\geq 4\sqrt{\sqrt{ab} \cdot \sqrt{cd}} = 4\sqrt[4]{abcd}.$$

Can we deal with three numbers in a similar way? Well, not quite, but we will present two ways to prove the analogous inequality for three variables: one way which generalizes and one which doesn't. We start with the one that doesn't. Let a, b, c be nonnegative real numbers. The identity
$$a+b+c-3\sqrt[3]{abc} = (\sqrt[3]{a} + \sqrt[3]{b} + \sqrt[3]{c}) \cdot \frac{(\sqrt[3]{a} - \sqrt[3]{b})^2 + (\sqrt[3]{b} - \sqrt[3]{c})^2 + (\sqrt[3]{c} - \sqrt[3]{a})^2}{2}$$
is obtained from the classical identity
$$x^3 + y^3 + z^3 - 3xyz = (x + y + z)\frac{(x - y)^2 + (y - z)^2 + (z - x)^2}{2}$$
by substituting $x = \sqrt[3]{a}$, $y = \sqrt[3]{b}$ and $z = \sqrt[3]{c}$. The previous relation clearly shows that
$$a + b + c \geq 3\sqrt[3]{abc}.$$

Let us turn to the second method (which is based on a wonderful idea of Cauchy): take the already established inequality
$$a + b + c + d \geq 4\sqrt[4]{abcd}$$
and choose $d = \frac{a+b+c}{3}$. We obtain
$$\frac{4}{3}(a + b + c) \geq 4\sqrt[4]{abc\frac{a + b + c}{3}}.$$

Dividing by 4, raising this to the fourth power and finally dividing by $a+b+c$ yields
$$(a+b+c)^3 \geq 27abc,$$
which is equivalent to
$$a+b+c \geq 3\sqrt[3]{abc}.$$

We will imitate this argument in the general case to prove the very important

Theorem 6.1. *(AM-GM inequality) For all nonnegative real numbers $x_1, x_2, ..., x_n$ we have*
$$\frac{x_1 + x_2 + ... + x_n}{n} \geq \sqrt[n]{x_1 x_2 ... x_n}.$$

Thus the geometric mean of some nonnegative numbers never exceeds their arithmetic mean.

Solution. We start by proving by induction on k that
$$x_1 + x_2 + ... + x_{2^k} \geq 2^k \sqrt[2^k]{x_1 x_2 ... x_{2^k}}$$

for all nonnegative real numbers $x_1, ..., x_{2^k}$ and all $k \geq 1$. The case $k = 1$ has already been seen, and to pass from k to $k+1$ simply apply twice the inductive hypothesis:
$$x_1 + x_2 + ... + x_{2^k} + ... + x_{2^{k+1}} \geq 2^k \sqrt[2^k]{x_1 ... x_{2^k}} + 2^k \sqrt[2^k]{x_{2^k+1} ... x_{2^{k+1}}}$$
$$\geq 2^{k+1} \sqrt[2^{k+1}]{x_1 ... x_{2^{k+1}}},$$

the last inequality being simply the case $k = 1$ applied to $\sqrt[2^k]{x_1 ... x_{2^k}}$ and $\sqrt[2^k]{x_{2^k+1} ... x_{2^{k+1}}}$. This establishes the inductive step.

Now let us take any positive integer n and nonnegative real numbers $x_1, ..., x_n$. Take k such that $2^k > n$ (for instance $k = n$) and set
$$x_{n+1} = ... = x_{2^k} = \frac{x_1 + ... + x_n}{n}.$$

Using the inequality we have already established, we obtain
$$x_1 + ... + x_n + (2^k - n)\frac{x_1 + ... + x_n}{n} \geq 2^k \sqrt[2^k]{x_1 ... x_n \left(\frac{x_1 + ... + x_n}{n}\right)^{2^k - n}}.$$

After division by $2^k \cdot \frac{x_1 + ... + x_n}{n}$ and raising the resulting inequality to the 2^kth power, this rather complicated expression becomes simply
$$1 \geq x_1 x_2 ... x_n \cdot \left(\frac{x_1 + ... + x_n}{n}\right)^{-n},$$

or equivalently
$$x_1 + x_2 + ... + x_n \geq n \sqrt[n]{x_1 x_2 ... x_n},$$
which is precisely the content of the theorem.

We can also prove that we have equality in the previous inequality if and only if $x_1 = ... = x_n$. We leave this as an easy exercise to the reader (who will of course need to go back to the proof presented above in order to establish the equality case!).

The following consequence of the AM-GM inequality is very important.

Theorem 6.2. *(AM-GM-HM inequality) For all positive numbers $a_1, a_2, ..., a_n$ we have*
$$\frac{a_1 + a_2 + ... + a_n}{n} \geq \sqrt[n]{a_1 a_2 ... a_n} \geq \frac{n}{\frac{1}{a_1} + \frac{1}{a_2} + ... + \frac{1}{a_n}}.$$

Proof. The inequality on the left is simply the AM-GM inequality, that we have already established. For the inequality on the right, writing it as
$$\frac{1}{a_1} + \frac{1}{a_2} + ... + \frac{1}{a_n} \geq n \sqrt[n]{\frac{1}{a_1} \cdot \frac{1}{a_2} \cdot ... \cdot \frac{1}{a_n}}$$
makes it clear that it is also a consequence of the AM-GM inequality. The result follows. $\square$

We end this theoretical part with another very useful inequality.

Theorem 6.3. *(Hölder's inequality) Let $a_{11}, ..., a_{1n}, a_{21}, ..., a_{kn}$ be nonnegative real numbers. Then*
$$(a_{11} + a_{12} + ... + a_{1n})(a_{21} + a_{22} + ... + a_{2n})...(a_{k1} + ... + a_{kn}) \geq$$
$$\left(\sqrt[k]{a_{11} a_{21} ... a_{k1}} + ... + \sqrt[k]{a_{1n} a_{2n} ... a_{kn}} \right)^k.$$

Proof. Let
$$S_1 = a_{11} + a_{12} + ... + a_{1n}, ..., S_k = a_{k1} + ... + a_{kn}.$$
Taking the kth root we obtain the equivalent inequality:
$$\sqrt[k]{a_{11} a_{21} ... a_{k1}} + ... + \sqrt[k]{a_{1n} a_{2n} ... a_{kn}} \leq \sqrt[k]{S_1 S_2 ... S_k}.$$
Dividing by the right hand-side, we write the inequality as
$$\sqrt[k]{\frac{a_{11} a_{21} ... a_{k1}}{S_1 ... S_k}} + ... + \sqrt[k]{\frac{a_{1n} a_{2n} ... a_{kn}}{S_1 ... S_k}} \leq 1.$$

AM-GM and Hölder's inequality

This follows by adding the following inequalities, obtained from the AM-GM inequality

$$\sqrt[k]{\frac{a_{11}a_{21}...a_{k1}}{S_1...S_k}} \leq \frac{1}{k}\left(\frac{a_{11}}{S_1} + ... + \frac{a_{k1}}{S_k}\right), ...,$$

$$\sqrt[k]{\frac{a_{1n}a_{2n}...a_{kn}}{S_1...S_k}} \leq \frac{1}{k}\left(\frac{a_{1n}}{S_1} + ... + \frac{a_{kn}}{S_k}\right).$$

□

Let us now see how to play with these inequalities in practice.

Example 6.1. Prove that for all $a, b, c \geq 0$ we have
 a) $(a+b)(b+c)(c+a) \geq 8abc$.
 b) $(a+b)(b+c)(c+a) \geq \frac{8}{9}(a+b+c)(ab+bc+ca)$.
 c) Which inequality is stronger?

Solution. a) This follows simply by multiplying the inequalities

$$a+b \geq 2\sqrt{ab}, \quad b+c \geq 2\sqrt{bc}, \quad c+a \geq 2\sqrt{ca}.$$

b) Let $S = a+b+c$. Then

$$(a+b)(b+c)(c+a) = (S-a)(S-b)(S-c) = S^3 - S^2(a+b+c) + S(ab+bc+ca) - abc$$

$$= S(ab+bc+ca) - abc.$$

Thus the inequality is equivalent to

$$S(ab+bc+ca) \geq 9abc \quad \text{or} \quad (a+b+c)\left(\frac{1}{a} + \frac{1}{b} + \frac{1}{c}\right) \geq 9.$$

But this last inequality is simply a reformulation of the AM-HM inequality. Note that we could also have expanded everything and obtained the equivalent inequality

$$a(b-c)^2 + b(c-a)^2 + c(a-b)^2 \geq 0,$$

which is clear.

c) The proof of b) shows that

$$\frac{8}{9}(a+b+c)(ab+bc+ca) \geq 8abc,$$

hence the inequality in b) is stronger than the one in a).

Example 6.2. Prove that for all $x, y > 0$

$$\frac{x}{x^4+y^2} + \frac{y}{y^4+x^2} \leq \frac{1}{xy}.$$

Solution. We use the AM-GM inequality

$$\frac{x}{x^4+y^2} \leq \frac{x}{2x^2y} = \frac{1}{2xy}$$

and similarly we obtain $\frac{y}{y^4+x^2} \leq \frac{1}{2xy}$. Adding these two inequalities yields the desired result.

Example 6.3. Let a and b be positive real numbers. Prove that

$$\left(\frac{a}{b}+\frac{b}{a}\right)^3 \geq 3\left(\frac{a}{b}+\frac{b}{a}\right)+2.$$

Solution. It is quite apparent that only $x = \frac{a}{b}+\frac{b}{a}$ plays a role in this problem. By the AM-GM inequality we have $x \geq 2$. It suffices therefore to prove the inequality $x^3 \geq 3x+2$ when $x \geq 2$. Let us factor

$$x^3 - 3x - 2 = x^3 - 4x + x - 2 = x(x^2-4) + x - 2$$
$$= (x-2)(x(x+2)+1) = (x-2)(x+1)^2 \geq 0.$$

The result follows.

Example 6.4. Let a, b, c be real numbers. Prove that

$$\frac{a^2-b^2}{2a^2+1} + \frac{b^2-c^2}{2b^2+1} + \frac{c^2-a^2}{2c^2+1} \leq 0.$$

Solution. An obvious change of variable would be $x = a^2$, $y = b^2$ and $z = c^2$, but an even better one is

$$x = 2a^2+1, \quad y = 2b^2+1, \quad z = 2c^2+1,$$

since then

$$a^2 - b^2 = \frac{x-y}{2}$$

and similarly for the other terms. Hence the inequality can be written as

$$\frac{x-y}{x} + \frac{y-z}{y} + \frac{z-x}{z} \leq 0,$$

or

$$\frac{y}{x} + \frac{z}{y} + \frac{x}{z} \geq 3,$$

which is clearly true by AM-GM.

Example 6.5. Let a, b, c be positive real numbers. Prove that

$$\frac{a^3}{(a+b)^2} + \frac{b^3}{(b+c)^2} + \frac{c^3}{(c+a)^2} \geq \frac{a+b+c}{4}.$$

Solution. This follows from Hölder's inequality, since

$$((a+b) + (b+c) + (c+a))^2 \cdot \left(\frac{a^3}{(a+b)^2} + \frac{b^3}{(b+c)^2} + \frac{c^3}{(c+a)^2} \right)$$

$$\geq \left(\sqrt[3]{(a+b)^2 \cdot \frac{a^3}{(a+b)^2}} + \sqrt[3]{(b+c)^2 \cdot \frac{b^3}{(b+c)^2}} + \sqrt[3]{(c+a)^2 \cdot \frac{c^3}{(c+a)^2}} \right)^3$$

$$= (a+b+c)^3.$$

Dividing by $4(a+b+c)^2$ yields the desired result.

Example 6.6. Let p and q be positive real numbers such that $\frac{1}{p} + \frac{1}{q} = 1$. Prove that

$$\frac{1}{(p-1)(q-1)} - \frac{1}{(p+1)(q+1)} \geq \frac{8}{9}.$$

Solution. Let us change the variables and denote $x = \frac{1}{p}$, $y = \frac{1}{q}$. Then the hypothesis becomes $x + y = 1$ and $x, y > 0$, and we need to prove that

$$\frac{xy}{(1-x)(1-y)} - \frac{xy}{(1+x)(1+y)} \geq \frac{8}{9}.$$

Since $1 - x = y$ and $1 - y = x$, the first term in the left hand-side equals 1, hence the inequality is equivalent to

$$\frac{xy}{(1+x)(1+y)} \leq \frac{1}{9}.$$

This can be rewritten as

$$1 + x + y + xy \geq 9xy, \quad \text{or} \quad xy \leq \frac{1}{4}.$$

But this follows from the AM-GM inequality and the hypothesis $x + y = 1$.

Example 6.7. If a, b, c are real numbers in $(0, 4)$, prove that at least one of the following numbers

$$\frac{1}{a} + \frac{1}{4-b}, \quad \frac{1}{b} + \frac{1}{4-c}, \quad \frac{1}{c} + \frac{1}{4-a}$$

is greater than or equal to 1.

Solution. In this kind of problems, reasoning by contradiction is always a good step. Let us suppose that the conclusion fails, i.e. that

$$\frac{1}{a} + \frac{1}{4-b} < 1, \quad \frac{1}{b} + \frac{1}{4-c} < 1, \quad \frac{1}{c} + \frac{1}{4-a} < 1.$$

We add these inequalities and we rearrange terms in order to separate the variables, obtaining
$$\frac{1}{a} + \frac{1}{4-a} + \frac{1}{b} + \frac{1}{4-b} + \frac{1}{c} + \frac{1}{4-c} < 3.$$
On the other hand, the AM-HM inequality yields
$$\frac{1}{x} + \frac{1}{4-x} \geq \frac{4}{x+(4-x)} = 1$$
for all $x \in (0,4)$. Writing these inequalities for $x = a, b, c$ and adding them yields the desired contradiction.

Example 6.8. Let $a, b > 0$ be such that
$$|a - 2b| \leq \frac{1}{\sqrt{a}} \quad \text{and} \quad |b - 2a| \leq \frac{1}{\sqrt{b}}.$$
Prove that $a + b \leq 2$.

Solution. Let us multiply the relations by $\sqrt{a}$ and $\sqrt{b}$ respectively and then take their squares. We obtain
$$a(a-2b)^2 \leq 1, \quad \text{and} \quad b(2a-b)^2 \leq 1.$$
Adding these two relations yields
$$a^3 - 4a^2b + 4ab^2 + 4a^2b - 4ab^2 + b^3 \leq 2,$$
that is $a^3 + b^3 \leq 2$. Using this relation and Hölder's inequality, we obtain
$$8 \geq 4(a^3 + b^3) = (1^3 + 1^3)(1^3 + 1^3)(a^3 + b^3) \geq (a+b)^3,$$
hence $a + b \leq 2$, as desired.

Example 6.9. Let a, b, c be positive real numbers such that $\frac{1}{a} + \frac{1}{b} + \frac{1}{c} = 1$. Prove that
$$\frac{1}{(a-1)(b-1)(c-1)} + \frac{8}{(a+1)(b+1)(c+1)} \leq \frac{1}{4}.$$
Solution. Let us make the substitution
$$x = \frac{1}{a}, \quad y = \frac{1}{b}, \quad z = \frac{1}{c}.$$
The hypothesis becomes $x + y + z = 1$ and the conclusion
$$\frac{xyz}{(1-x)(1-y)(1-z)} + \frac{8xyz}{(1+x)(1+y)(1+z)} \leq \frac{1}{4}.$$

Now,
$$(1-x)(1-y)(1-z) = (x+y)(y+z)(z+x) \geq 8xyz,$$

where the last inequality was established in exercise 1. It suffices therefore to prove that
$$\frac{8xyz}{(1+x)(1+y)(1+z)} \leq \frac{1}{8}.$$

This follows from
$$(1+x)(1+y)(1+z) = (2x+y+z)(2y+z+x)(2z+x+y) =$$
$$((x+y)+(x+z))((y+x)+(y+z))((z+x)+(z+y))$$
$$\geq 8(x+y)(y+z)(z+x) \geq 64xyz,$$

where we have used again exercise 1 with $a = y+z, b = z+x$ and $c = x+y$ and then with $a = x, b = y$ and $c = z$.

Example 6.10. Let a, b, c be real numbers greater than or equal to 1. Prove that
$$\frac{a^3}{b^2 - b + \frac{1}{3}} + \frac{b^3}{c^2 - c + \frac{1}{3}} + \frac{c^3}{a^2 - a + \frac{1}{3}} \geq 9.$$

Solution. Write the inequality as
$$\frac{a^3}{3b^2 - 3b + 1} + \frac{b^3}{3c^2 - 3c + 1} + \frac{c^3}{3a^2 - 3a + 1} \geq 3.$$

We recognize a few terms in the expansion of
$$(x-1)^3 = x^3 - 3x^2 + 3x - 1 = x^3 - (3x^2 - 3x + 1).$$

Hence we can rewrite the inequality as
$$\frac{a^3}{b^3 - (b-1)^3} + \frac{b^3}{c^3 - (c-1)^3} + \frac{c^3}{a^3 - (a-1)^3} \geq 3.$$

Since by hypothesis $a, b, c \geq 1$, we have $a^3 - (a-1)^3 \leq a^3$ and similar inequalities with b and c instead of a. Hence it suffices to prove that
$$\frac{a^3}{b^3} + \frac{b^3}{c^3} + \frac{c^3}{a^3} \geq 3,$$

which is a straight consequence of the AM-GM inequality.

Example 6.11. Let a, b, c be positive real numbers. Prove that
$$\frac{a^2}{b^3} + \frac{b^2}{c^3} + \frac{c^2}{a^3} \geq \frac{1}{a} + \frac{1}{b} + \frac{1}{c}.$$

Solution. This is a pretty tricky application of Hölder's inequality:

$$\left(\frac{a^2}{b^3}+\frac{b^2}{c^3}+\frac{c^2}{a^3}\right)\left(\frac{1}{a}+\frac{1}{b}+\frac{1}{c}\right)\left(\frac{1}{a}+\frac{1}{b}+\frac{1}{c}\right) \geq \left(\frac{1}{b}+\frac{1}{c}+\frac{1}{a}\right)^3,$$

and the result follows by dividing by $\left(\frac{1}{a}+\frac{1}{b}+\frac{1}{c}\right)^2$.

Example 6.12. Prove that for all $a, b, c, d > 0$

$$\frac{a+c}{a+b}+\frac{b+d}{b+c}+\frac{a+c}{c+d}+\frac{b+d}{a+d} \geq 4.$$

Solution. We pair the first and third terms, as well as the second and fourth terms and we obtain

$$\frac{a+c}{a+b}+\frac{b+d}{b+c}+\frac{a+c}{c+d}+\frac{b+d}{a+c} = (a+c)\left(\frac{1}{a+b}+\frac{1}{c+d}\right)+(b+d)\left(\frac{1}{b+c}+\frac{1}{a+d}\right).$$

Using the AM-HM inequality, we can write

$$\frac{1}{a+b}+\frac{1}{c+d} \geq \frac{4}{a+b+c+d}, \quad \frac{1}{b+c}+\frac{1}{a+d} \geq \frac{4}{a+b+c+d}.$$

Hence

$$\frac{a+c}{a+b}+\frac{b+d}{b+c}+\frac{a+c}{c+d}+\frac{b+d}{a+c} \geq \frac{4(a+c)}{a+b+c+d}+\frac{4(b+d)}{a+b+c+d} = 4$$

and we are done.

Example 6.13. Let x, y, z be positive real numbers such that

$$xy+yz+zx \geq \frac{1}{\sqrt{x^2+y^2+z^2}}.$$

Prove that $x+y+z \geq \sqrt{3}$.

Solution. This is quite tricky. Write the hypothesis in the simpler form

$$(xy+yz+zx)^2(x^2+y^2+z^2) \geq 1$$

by squaring and cross multiplying. On the other hand, by the AM-GM inequality we obtain

$$(xy+yz+zx)^2(x^2+y^2+z^2) \leq \left(\frac{2(xy+yz+zx)+x^2+y^2+z^2}{3}\right)^3$$

$$= \left(\frac{(x+y+z)^2}{3}\right)^3.$$

Combining the two inequalities yields $\frac{(x+y+z)^2}{3} \geq 1$ and then $x+y+z \geq \sqrt{3}$.

Example 6.14. Let a, b, c be positive real numbers such that $a + b + c = 1$. Prove that
$$\left(1 + \frac{1}{a}\right)\left(1 + \frac{1}{b}\right)\left(1 + \frac{1}{c}\right) \geq 64.$$

Solution. The shortest solution is based on Hölder's inequality:
$$\left(1 + \frac{1}{a}\right)\left(1 + \frac{1}{b}\right)\left(1 + \frac{1}{c}\right) \geq \left(1 + \sqrt[3]{\frac{1}{abc}}\right)^3,$$

hence it suffices to prove that the last quantity is at least 64. But this comes down to $abc \leq \frac{1}{27}$, which is a consequence of the AM-GM inequality and the hypothesis $a + b + c = 1$.

For an alternative solution, let us start with a brutal expansion
$$\left(1 + \frac{1}{a}\right)\left(1 + \frac{1}{b}\right)\left(1 + \frac{1}{c}\right) = 1 + \frac{1}{a} + \frac{1}{b} + \frac{1}{c} + \frac{1}{ab} + \frac{1}{ac} + \frac{1}{bc} + \frac{1}{abc}$$
$$= 1 + \frac{1}{a} + \frac{1}{b} + \frac{1}{c} + \frac{a+b+c}{abc} + \frac{1}{abc} = 1 + \frac{1}{a} + \frac{1}{b} + \frac{1}{c} + \frac{2}{abc},$$

the last equality being a consequence of the hypothesis $a + b + c = 1$. We will bound from below separately $\frac{1}{abc}$ and $\frac{1}{a} + \frac{1}{b} + \frac{1}{c}$. Bounding $\frac{1}{abc}$ from below comes down to bounding abc from above. This is quite easy since the hypothesis $a + b + c = 1$ yields, via AM-GM
$$1 = a + b + c \geq 3\sqrt[3]{abc}.$$

Thus $abc \leq \frac{1}{27}$ and $\frac{2}{abc} \geq 54$. Thus, it suffices to prove that
$$\frac{1}{a} + \frac{1}{b} + \frac{1}{c} \geq 9$$

to conclude the proof of the original inequality. This is a consequence of the AM-HM inequality
$$\frac{1}{a} + \frac{1}{b} + \frac{1}{c} \geq \frac{3^2}{a+b+c} = 9$$
or a consequence of the AM-GM inequality
$$\frac{1}{a} + \frac{1}{b} + \frac{1}{c} \geq 3\sqrt[3]{\frac{1}{abc}}$$

combined with $abc \leq \frac{1}{27}$, which has already been proved.

Example 6.15. Prove that for all positive real numbers a, b, c
$$\left(1 + \frac{a}{b}\right)\left(1 + \frac{b}{c}\right)\left(1 + \frac{c}{a}\right) \geq 2\left(1 + \frac{a+b+c}{\sqrt[3]{abc}}\right).$$

Solution. Expanding, we obtain the equivalent inequality

$$\frac{a}{b} + \frac{b}{c} + \frac{c}{a} + \frac{b}{a} + \frac{c}{b} + \frac{a}{c} \geq 2\frac{a+b+c}{\sqrt[3]{abc}}.$$

This is a homogeneous inequality, so we may assume that $abc = 1$. We will prove that

$$\frac{a}{b} + \frac{b}{c} + \frac{c}{a} \geq a+b+c \quad \text{and} \quad \frac{b}{a} + \frac{c}{b} + \frac{a}{c} \geq a+b+c,$$

which is enough to conclude. By permuting the variables a, b, c, it suffices to prove the first inequality. Note that

$$\frac{2a}{b} + \frac{b}{c} = \frac{a}{b} + \frac{a}{b} + \frac{b}{c} \geq 3\sqrt[3]{\frac{a^2}{bc}} = 3a,$$

by the AM-GM inequality and the hypothesis $abc = 1$. Similar arguments yield

$$\frac{2b}{c} + \frac{c}{a} \geq 3b, \quad \frac{2c}{a} + \frac{a}{b} \geq 3c.$$

Adding up these inequalities yields the desired result.

Example 6.16. Prove that if a, b, c are nonnegative real numbers such that $a + b + c = 3$, then

$$abc(a^2 + b^2 + c^2) \leq 3.$$

Solution. The inequality

$$(ab + bc + ca)^2 \geq 3abc(a + b + c)$$

is equivalent after expansion to

$$(ab - bc)^2 + (bc - ca)^2 + (ca - ab)^2 \geq 0$$

and thus true. Combined with the hypothesis, we obtain therefore

$$abc \leq \frac{(ab + bc + ca)^2}{9}.$$

Hence it suffices to prove the stronger inequality

$$(ab + bc + ca)^2(a^2 + b^2 + c^2) \leq 27.$$

But this follows from the AM-GM inequality

$$(ab + bc + ca)^2(a^2 + b^2 + c^2) \leq \left(\frac{a^2 + b^2 + c^2 + 2(ab + bc + ca)}{3}\right)^3 = 27,$$

the last equality being a consequence of the hypothesis and of the identity

$$a^2 + b^2 + c^2 + 2(ab + bc + ca) = (a + b + c)^2.$$

Example 6.17. Prove that for all $a, b, c \geq 0$
$$(a^2 + ab + b^2)(b^2 + bc + c^2)(c^2 + ca + a^2) \geq (ab + bc + ca)^3.$$

Solution. This is an immediate consequence of Hölder's inequality:
$$(ab + b^2 + a^2)(b^2 + bc + c^2)(a^2 + c^2 + ac) \geq (ab + bc + ca)^3.$$

If you found this tricky, here is an alternative approach. We have
$$a^2 + b^2 + ab = (a+b)^2 - ab \geq (a+b)^2 - \frac{(a+b)^2}{4} = \frac{3}{4}(a+b)^2,$$
hence it suffices to prove the stronger inequality
$$\frac{27}{64}(a+b)^2(b+c)^2(c+a)^2 \geq (ab+bc+ca)^3.$$

Using the inequality
$$(a+b)(b+c)(c+a) \geq \frac{8}{9}(a+b+c)(ab+bc+ca)$$

of exercise 1, it suffices to prove that
$$(a+b+c)^2 \geq 3(ab+bc+ca)$$

which after expansion is equivalent to
$$(a-b)^2 + (b-c)^2 + (c-a)^2 \geq 0$$

and thus true.

Example 6.18. Prove that for all positive real numbers a, b, c the following inequality holds
$$(a^5 - a^2 + 3)(b^5 - b^2 + 3)(c^5 - c^2 + 3) \geq (a+b+c)^3.$$

Solution. This is fairly tricky. We start by using Hölder's inequality in the form
$$(a^3 + 1^3 + 1^3)(b^3 + 1^3 + 1^3)(c^3 + 1^3 + 1^3) \geq (a+b+c)^3,$$

hence it suffices to prove that
$$(a^5 - a^2 + 3)(b^5 - b^2 + 3)(c^5 - c^2 + 3) \geq (a^3 + 2)(b^3 + 2)(c^3 + 2).$$

The advantage is that this is an inequality with separated variables, so it suffices to prove that
$$x^5 - x^2 + 3 \geq x^3 + 2$$

for all $x \geq 0$. This can be written as
$$x^5 - x^3 - x^2 + 1 \geq 0 \quad \text{or} \quad (x^2 - 1)(x^3 - 1) \geq 0$$

or finally as
$$(x-1)^2(x+1)(x^2 + x + 1) \geq 0,$$

making it clear.

7 Lagrange's identity and the Cauchy-Schwarz inequality

Let us start with an easy case: rewriting $(a^2 + b^2)(c^2 + d^2)$ as the sum of two squares. In order to do that, we expand

$$(a^2 + b^2)(c^2 + d^2) = (ac)^2 + (ad)^2 + (bc)^2 + (bd)^2.$$

Next, we add and subtract $2abcd$, in order to complete the square

$$(ac)^2 + (ad)^2 + (bc)^2 + (bd)^2 = (ac)^2 + 2acbd + (bd)^2 + (ad)^2 + (bc)^2 - 2abcd =$$
$$(ac + bd)^2 + (ad - bc)^2$$

and we obtain the special case

$$(a^2 + b^2)(c^2 + d^2) = (ac + bd)^2 + (ad - bc)^2$$

of a general and very useful result:

Theorem 7.1. *(Lagrange's identity) For all real numbers* $x_1, x_2, ..., x_n$ *and* $y_1, y_2, ..., y_n$ *we have*

$$(x_1^2 + x_2^2 + ... + x_n^2)(y_1^2 + y_2^2 + ... + y_n^2)$$
$$= (x_1y_1 + x_2y_2 + ... + x_ny_n)^2 + \sum_{1 \le i < j \le n}(x_iy_j - x_jy_i)^2.$$

Proof. Using the formula

$$(\sum_{i=1}^{n} x_i) \cdot (\sum_{j=1}^{m} y_j) = \sum_{i=1}^{n}\sum_{j=1}^{m} x_i y_j,$$

we can write the left hand-side as

$$(x_1^2 + x_2^2 + ... + x_n^2)(y_1^2 + y_2^2 + ... + y_n^2) = \sum_{i=1}^{n}\sum_{j=1}^{n}(x_iy_j)^2.$$

We separate the terms in the above sum according to whether $i > j$, $i < j$ or $i = j$. Thus

$$\sum_{i=1}^{n}\sum_{j=1}^{n}(x_iy_j)^2 = \sum_{i=1}^{n}(x_iy_i)^2 + \sum_{1 \le i < j \le n}((x_iy_j)^2 + (x_jy_i)^2).$$

We complete the square in each parenthesis of the sum indexed by $i < j$, by adding and subtracting $2x_ix_jy_iy_j$. We end up with

$$\sum_{1 \le i < j \le n}((x_iy_j)^2 + (x_jy_i)^2) = \sum_{1 \le i < j \le n}(x_iy_j - x_jy_i)^2 + 2\sum_{1 \le i < j \le n} x_iy_ix_jy_j,$$

which combined with the previous relations yields

$$(x_1^2+...+x_n^2)(y_1^2+...+y_n^2) = \sum_{i=1}^{n}(x_iy_i)^2 + 2\sum_{1\leq i<j\leq n} x_iy_ix_jy_j + \sum_{1\leq i<j\leq n}(x_iy_j-x_jy_i)^2.$$

We conclude using the identity

$$\sum_{i=1}^{n} z_i^2 + 2\sum_{1\leq i<j\leq n} z_iz_j = \left(\sum_{i=1}^{n} z_i\right)^2,$$

applied to $z_i = x_iy_i$. □

An immediate consequence of Lagrange's identity is the following extremely useful inequality

Theorem 7.2. *(Cauchy-Schwarz inequality) For all real numbers $x_1, x_2, ..., x_n$ and $y_1, y_2, ..., y_n$ we have*

$$(x_1^2 + x_2^2 + ... + x_n^2)(y_1^2 + y_2^2 + ... + y_n^2) \geq (x_1y_1 + x_2y_2 + ... + x_ny_n)^2,$$

with equality if and only if either $x_i = 0$ for all i or there is a real number t such that $y_i = tx_i$ for all $i = 1, ..., n$.

Proof. The first part is a direct consequence of Lagrange's identity, which expresses the difference between the left hand-side and the right hand-side as a sum of squares. Again by Lagrange's identity, we have equality if and only if $x_iy_j = x_jy_i$ for all $i < j$, or equivalently for all i and j (since by symmetry this continues to hold for $i > j$ and since the equality is clear for $i = j$). We need to prove that this happens if and only if either $x_i = 0$ for all i or we can find t such that $y_i = tx_i$ for all i. One implication being clear, let us try to prove the converse. So assume that $x_iy_j = x_jy_i$ for all i, j. We would like to write the hypothesis as $\frac{y_i}{x_i} = \frac{y_j}{x_j}$, which would make the conclusion clear. However, it may happen that some x_i are equal to 0. Let us discuss two cases. First, if all x_i are zero, then we are done. Otherwise, pick some i_0 for which $x_{i_0} \neq 0$ and set $t = \frac{y_{i_0}}{x_{i_0}}$. Then

$$tx_i = \frac{y_{i_0}x_i}{x_{i_0}} = \frac{y_ix_{i_0}}{x_{i_0}} = y_i$$

for all i and we are done. □

Let us give a different proof of the Cauchy-Schwarz inequality, based on the theory of quadratic equations: let us introduce a new variable t and consider

$$f(t) = (tx_1 - y_1)^2 + ... + (tx_n - y_n)^2.$$

We will exclude from now on the case when all x_i are 0, since in this case the conclusion is clear. Then $f(t)$ is a quadratic polynomial in t, since expanding and rearranging terms gives

$$f(t) = t^2 x_1^2 - 2t x_1 y_1 + y_1^2 + \ldots + t^2 x_n^2 - 2t x_n y_n + y_n^2 =$$
$$t^2(x_1^2 + \ldots + x_n^2) - 2t(x_1 y_1 + \ldots + x_n y_n) + y_1^2 + \ldots + y_n^2.$$

Since $f(t)$ is a sum of squares, it is nonnegative for all values of t, thus its discriminant Δ must be nonpositive. But the previous expression of f shows that

$$\Delta = 4((x_1 y_1 + \ldots + x_n y_n)^2 - (x_1^2 + \ldots + x_n^2)(y_1^2 + \ldots + y_n^2)).$$

The inequality follows. To establish the equality case, note that if equality holds, then $\Delta = 0$, hence the equation $f(t) = 0$ must have exactly one real solution t. But this equation can be written as $(tx_1 - y_1)^2 + \ldots + (tx_n - y_n)^2 = 0$, hence we must have $tx_i = y_i$ for all i.

A very important special case of the Cauchy-Schwarz inequality is obtained by taking $y_1 = \ldots = y_n = 1$. We obtain the inequality

$$x_1^2 + x_2^2 + \ldots + x_n^2 \geq \frac{(x_1 + x_2 + \ldots + x_n)^2}{n},$$

which is extremely useful when trying to bound from below sums of squares. For instance, the case $n = 2$ is

$$a^2 + b^2 \geq \frac{(a+b)^2}{2},$$

which is itself a very useful inequality (equivalent to the fundamental inequality $(a-b)^2 \geq 0$). Also, note that the previous inequality can be written as

$$\sqrt{\frac{x_1^2 + x_2^2 + \ldots + x_n^2}{n}} \geq \frac{x_1 + x_2 + \ldots + x_n}{n},$$

that is the quadratic mean of some real numbers is always greater than or equal to their arithmetic mean!

Theorem 7.3. *(QM-AM-HM inequality) For all positive real numbers $x_1, x_2, \ldots, x_n$ we have*

$$\sqrt{\frac{x_1^2 + x_2^2 + \ldots + x_n^2}{n}} \geq \frac{x_1 + x_2 + \ldots + x_n}{n} \geq \frac{n}{\frac{1}{x_1} + \frac{1}{x_n} + \ldots + \frac{1}{x_n}}.$$

In particular, for all $x_1, x_2, \ldots, x_n > 0$ we have

$$\frac{1}{x_1} + \frac{1}{x_n} + \ldots + \frac{1}{x_n} \geq \frac{n^2}{x_1 + x_2 + \ldots + x_n}.$$

Lagrange's identity and the Cauchy-Schwarz inequality

Proof. The second part of the theorem is simply a reformulation of the inequality on the right. Also, the inequality on the left has already been established in the discussion preceding the theorem. The inequality on the right was established in the chapter on AM-GM and Hölder's inequality, but we can also give a proof using the Cauchy-Schwarz inequality, as follows:

$$(x_1+x_2+\ldots+x_n)\left(\frac{1}{x_1}+\ldots+\frac{1}{x_n}\right) = (\sqrt{x_1}^2+\ldots+\sqrt{x_n}^2)\left(\sqrt{\frac{1}{x_1}}^2+\ldots+\sqrt{\frac{1}{x_n}}^2\right)$$

$$\geq (1+\ldots+1)^2 = n^2.$$

$\square$

We can generalize the second part of the previous theorem and this is the content of the next direct corollary of the Cauchy-Schwarz inequality. This corollary is the key ingredient for a host of olympiad problems!

Corollary 7.4. *For all $a_1, a_2, \ldots, a_n \in \mathbf{R}$ and all $b_1, b_2, \ldots, b_n > 0$ we have*

$$\frac{a_1^2}{b_1} + \frac{a_2^2}{b_2} + \ldots + \frac{a_n^2}{b_n} \geq \frac{(a_1 + a_2 + \ldots + a_n)^2}{b_1 + b_2 + \ldots + b_n}.$$

Proof. The Cauchy-Schwarz inequality gives

$$\left(\frac{a_1^2}{b_1} + \ldots + \frac{a_n^2}{b_n}\right) \cdot (b_1 + \ldots + b_n) \geq \left(\sqrt{b_1 \cdot \frac{a_1^2}{b_1}} + \ldots + \sqrt{b_n \cdot \frac{a_n^2}{b_n}}\right)^2 =$$

$$(|a_1| + \ldots + |a_n|)^2 \geq |a_1 + \ldots + a_n|^2 = (a_1 + \ldots + a_n)^2,$$

where we have also used the triangle inequality

$$|a_1 + a_2 + \ldots + a_n| \leq |a_1| + \ldots + |a_n|$$

for all real numbers $a_1, \ldots, a_n$. Dividing by $b_1 + b_2 + \ldots + b_n$ yields the desired result.

$\square$

We insist on the fact that the denominators b_i should be **positive** in order to apply the previous corollary, while the a_i's can be any real numbers. We end up this theoretical part with another fundamental inequality that can be easily deduced from the Cauchy-Schwarz inequality.

Theorem 7.5. *(Minkowski's inequality) For all real numbers $x_1, x_2, \ldots, x_n$ and $y_1, y_2, \ldots, y_n$ the following inequality holds*

$$\sqrt{x_1^2 + y_1^2} + \sqrt{x_2^2 + y_2^2} + \ldots + \sqrt{x_n^2 + y_n^2}$$
$$\geq \sqrt{(x_1 + x_2 + \ldots + x_n)^2 + (y_1 + y_2 + \ldots + y_n)^2}.$$

Proof. Take the square of the desired inequality and use the identity

$$(z_1 + \ldots + z_n)^2 = z_1^2 + \ldots + z_n^2 + 2 \sum_{i<j} z_i z_j$$

to obtain the equivalent inequality

$$\sum_{i=1}^n (x_i^2 + y_i^2) + 2 \sum_{i<j} \sqrt{(x_i^2 + y_i^2)(x_j^2 + y_j^2)} \geq \sum_{i=1}^n (x_i^2 + y_i^2) + 2 \sum_{i<j} (x_i x_j + y_i y_j).$$

This simplifies quite nicely to

$$\sum_{i<j} \sqrt{(x_i^2 + y_i^2)(x_j^2 + y_j^2)} \geq \sum_{i<j} (x_i x_j + y_i y_j).$$

And this follows from the Cauchy-Schwarz inequality, since for all $i < j$ we have

$$(x_i x_j + y_i y_j)^2 \leq (x_i^2 + y_i^2)(x_j^2 + y_j^2),$$

hence

$$\sqrt{(x_i^2 + y_i^2)(x_j^2 + y_j^2)} \geq x_i x_j + y_i y_j.$$

□

We turn now to a long list of examples showing how to play with these inequalities in real life.

Example 7.1. Prove that for all real numbers a and b,

$$a^4 + b^4 \geq ab(a^2 + b^2).$$

Solution. Since $x^2 + y^2 \geq \frac{(x+y)^2}{2}$ and $\frac{x^2+y^2}{2} \geq xy$, we have

$$a^4 + b^4 \geq \frac{(a^2 + b^2)^2}{2} = \frac{a^2 + b^2}{2} \cdot (a^2 + b^2) \geq ab(a^2 + b^2)$$

and we are done.

Example 7.2. Prove that

$$\frac{(a+b)^2}{c} + \frac{c^2}{a} \geq 4b$$

for all $a, b, c > 0$.

Solution. By Cauchy-Schwarz (as in corollary 7.4), the left hand-side can be bounded from below by $\frac{(a+b+c)^2}{c+a}$, thus it suffices to prove that

$$\frac{(a+b+c)^2}{c+a} \geq 4b,$$

which comes down to $(a+b+c)^2 \geq 4b(a+c)$ and then to $(a+c-b)^2 \geq 0$.

Note that we can also prove the desired inequality using the AM-GM inequality twice:

$$a + b = a + 3 \cdot \frac{b}{3} \geq 4a^{1/4}(b/3)^{3/4}$$

and

$$\frac{(a+b)^2}{c} + \frac{c^2}{a} \geq 2\frac{8a^{1/2}b^{3/2}}{3^{3/2}c} + \frac{c^2}{a} \geq 3\left(\frac{8a^{1/2}b^{3/2}}{3^{3/2}c}\right)^{2/3}\left(\frac{c^2}{a}\right)^{1/3} = 4b.$$

Example 7.3. Prove that for all real numbers a and b,

$$4a^4 + 4b^4 \geq (a^2+b^2)(a+b)^2.$$

Solution. We have $(a+b)^2 \leq 2(a^2+b^2)$ by the Cauchy-Schwarz inequality, thus it suffices to prove that

$$4(a^4+b^4) \geq 2(a^2+b^2)^2$$

or equivalently

$$(a^2+b^2)^2 \leq 2(a^4+b^4).$$

This is another application of the Cauchy-Schwarz inequality.

Example 7.4. Let a, b, c, d be real numbers such that $a^2 + b^2 = c^2 + d^2 = 85$ and $|ac - bd| = 40$. Find $|ad + bc|$.

Solution. This is a rather easy consequence of Lagrange's identity

$$(ad+bc)^2 + (ac-bd)^2 = (a^2+b^2)(c^2+d^2).$$

We obtain

$$(ad+bc)^2 = 85^2 - 40^2 = (85-40)(85+40) = 45 \cdot 125 = 9 \cdot 5 \cdot 5^3 = (3 \cdot 5^2)^2,$$

hence $|ad + bc| = 3 \cdot 5^2 = 75$.

Example 7.5. Let a and b be positive real numbers. Prove that

$$(a+b)^3 \leq 4(a^3+b^3).$$

Solution. There are many ways to prove this inequality. One way is based on the fact that
$$a^3 + b^3 = (a+b)(a^2 - ab + b^2).$$
Since $ab \leq \frac{a^2+b^2}{2}$ and $a^2 + b^2 \geq \frac{(a+b)^2}{2}$, we obtain
$$a^3 + b^3 \geq (a+b)\frac{a^2+b^2}{2} \geq (a+b)\frac{(a+b)^2}{4} = \frac{(a+b)^3}{4}$$
and we are done.

We can also prove the inequality using Hölder's inequality:
$$4(a^3 + b^3) = (1^3 + 1^3)(1^3 + 1^3)(a^3 + b^3) \geq (a+b)^3.$$

Example 7.6. Let a and b be positive real numbers. Prove that
$$\frac{a+b}{2}\sqrt{\frac{a^2+b^2}{2}} \leq a^2 - ab + b^2.$$

Solution. With the same arguments as above, we have the chain of inequalities
$$a^2 - ab + b^2 \geq \frac{a^2+b^2}{2} = \sqrt{\frac{a^2+b^2}{2}} \cdot \sqrt{\frac{a^2+b^2}{2}} \geq \frac{a+b}{2}\sqrt{\frac{a^2+b^2}{2}}.$$

Example 7.7. Let a, b, c be positive real numbers such that $a+b+c = 1$. Prove that
$$\sqrt{9a+1} + \sqrt{9b+1} + \sqrt{9c+1} \leq 6.$$

Solution. This follows directly from the Cauchy-Schwarz inequality:
$$(\sqrt{9a+1} + \sqrt{9b+1} + \sqrt{9c+1})^2 \leq (1^2 + 1^2 + 1^2)(9a+1+9b+1+9c+1).$$

Since $a+b+c = 1$, the right hand-side equals $3 \cdot 12 = 36$. Taking the square root of the previous inequality yields the desired result.

Example 7.8. Let x, y be real numbers, not both 0. Prove that
$$\frac{x+y}{x^2 - xy + y^2} \leq \frac{2\sqrt{2}}{\sqrt{x^2+y^2}}.$$

Solution. First, observe that
$$x^2 - xy + y^2 = \left(x - \frac{y}{2}\right)^2 + \frac{3}{4}y^2 > 0,$$

since x, y are not both 0. Thus if $x + y < 0$, then the inequality is clear. Suppose that $x + y \geq 0$. Since $xy \leq \frac{x^2+y^2}{2}$, we have

$$x^2 - xy + y^2 \geq \frac{x^2 + y^2}{2},$$

hence

$$\frac{x+y}{x^2 - xy + y^2} \leq \frac{2(x+y)}{x^2 + y^2}.$$

It suffices therefore to prove that

$$\frac{x+y}{x^2+y^2} \leq \frac{\sqrt{2}}{\sqrt{x^2+y^2}}$$

or equivalently $(x+y)^2 \leq 2(x^2+y^2)$. This is equivalent to $(x-y)^2 \geq 0$, hence the result follows.

Example 7.9. If a, b, c, d, e are real numbers such that $-2 \leq a \leq b \leq c \leq d \leq e \leq 2$, prove that

$$\frac{1}{b-a} + \frac{1}{c-b} + \frac{1}{d-c} + \frac{1}{e-d} \geq 4.$$

Solution. We apply the AM-HM inequality to obtain

$$\frac{1}{b-a} + \frac{1}{c-b} + \frac{1}{d-c} + \frac{1}{e-d} \geq \frac{4^2}{b-a+c-b+d-c+e-d} = \frac{16}{e-a}.$$

It suffices therefore to prove that $\frac{16}{e-a} \geq 4$, or equivalently $e - a \leq 4$. This is clear, since $e - a \leq 2 - (-2) = 4$, by hypothesis.

Example 7.10. Find all triples (a, b, c) of positive real numbers such that

$$2\sqrt{a} + 3\sqrt{b} + 6\sqrt{c} = a + b + c = 49$$

Solution. Since we have only two equations and three variables, it is likely that some inequality is hidden in this problem. Using the Cauchy-Schwarz inequality, we can write

$$49^2 = (2\sqrt{a} + 3\sqrt{b} + 6\sqrt{c})^2 \leq (2^2 + 3^2 + 6^2)(a+b+c) = 49(a+b+c) = 49^2.$$

Thus we must have equality in the Cauchy-Schwarz inequality, which means that $(2, 3, 6)$ and $\sqrt{a}, \sqrt{b}, \sqrt{c}$ are proportional. Say $\sqrt{a} = 2x$, $\sqrt{b} = 3x$ and $\sqrt{c} = 6x$. Then the first equation of the system becomes $4x + 9x + 36x = 49$, thus $x = 1$ and then $a = 4$, $b = 9$ and $c = 36$. It is easy to check that this is indeed a solution of the problem, so the answer is $(a, b, c) = (4, 9, 36)$.

Note that we can avoid the use of the Cauchy-Schwarz inequality by considering the equality
$$a+b+c - 2(2\sqrt{a} + 3\sqrt{b} + 6\sqrt{c}) = 49 - 2 \cdot 49 = -49.$$
Separating variables and completing the square in the left hand-side yields the equivalent relation
$$(\sqrt{a} - 2)^2 + (\sqrt{b} - 3)^2 + (\sqrt{c} - 6)^2 = 0,$$
which immediately gives the unique solution $(a,b,c) = (4, 9, 36)$.

Example 7.11. Prove that in a triangle with side-lengths a, b, c,
$$\frac{1}{-a+b+c} + \frac{1}{a-b+c} + \frac{1}{a+b-c} \geq \frac{1}{a} + \frac{1}{b} + \frac{1}{c}.$$

Solution. A standard change of variables when dealing with sides of a triangle is
$$x = b+c-a, \quad y = c+a-b, \quad z = a+b-c.$$
Solving this linear system in a, b, c gives
$$a = \frac{y+z}{2}, \quad b = \frac{x+z}{2}, \quad c = \frac{x+y}{2}.$$
Replacing these in the original inequality, we obtain the equivalent inequality
$$\frac{1}{2x} + \frac{1}{2y} + \frac{1}{2z} \geq \frac{1}{x+y} + \frac{1}{x+z} + \frac{1}{z+y}.$$
This is an easy consequence of the AH-HM inequality:
$$\frac{1}{x} + \frac{1}{y} \geq \frac{2^2}{x+y}$$
and similar inequalities obtained by permuting x, y, z. Adding these inequalities yields the desired result.

Example 7.12. Prove that for all real numbers a, b, c
$$(1+a^2)(1+b^2)(1+c^2) = (a+b+c-abc)^2 + (ab+bc+ca-1)^2.$$

Solution. We apply Lagrange's identity twice:
$$(1+a^2)(1+b^2) = (a+b)^2 + (1-ab)^2$$
and
$$(1+a^2)(1+b^2)(1+c^2) = ((a+b)^2 + (1-ab)^2)(1^2+c^2) =$$
$$(a+b+c(1-ab))^2 + ((a+b)c-(1-ab))^2 = (a+b+c-abc)^2 + (ab+bc+ca-1)^2.$$

Lagrange's identity and the Cauchy-Schwarz inequality

Example 7.13. Let a, b, c, d be positive real numbers. Prove that
$$\frac{a}{a+2b+c} + \frac{b}{b+2c+d} + \frac{c}{c+2d+a} + \frac{d}{d+2a+b} \geq 1.$$

Solution. Using the Cauchy-Schwarz inequality in the form of corollary 7.4, we obtain
$$\sum \frac{a}{a+2b+c} = \sum \frac{a^2}{a^2+2ab+ac} \geq$$
$$\frac{(a+b+c+d)^2}{a^2+2ab+ac+b^2+2bc+bd+c^2+2cd+ca+d^2+2da+db}.$$

Next,
$$a^2+2ab+ac+b^2+2bc+bd+c^2+2cd+ca+d^2+2da+db =$$
$$\sum a^2 + 2\sum ab = (a+b+c+d)^2,$$
and the result follows.

Example 7.14. If a, b, c, d are positive real numbers such that $a+b+c+d = 8$, prove that
$$\left(a+\frac{1}{a}\right)^2 + \left(b+\frac{1}{b}\right)^2 + \left(c+\frac{1}{c}\right)^2 + \left(d+\frac{1}{d}\right)^2 \geq 25$$

Solution. There are many ways to solve this problem, but probably the cleanest one is the following, based on two simple applications of the QM-AM-HM inequality:
$$\left(a+\frac{1}{a}\right)^2 + \left(b+\frac{1}{b}\right)^2 + \left(c+\frac{1}{c}\right)^2 + \left(d+\frac{1}{d}\right)^2 \geq$$
$$\frac{1}{4}\left(a+\frac{1}{a}+b+\frac{1}{b}+c+\frac{1}{c}+d+\frac{1}{d}\right)^2 = \frac{1}{4}\left(8+\frac{1}{a}+\frac{1}{b}+\frac{1}{c}+\frac{1}{d}\right)^2,$$
the last equality being a consequence of the hypothesis $a+b+c+d = 8$. Hence, it suffices to prove that
$$\frac{1}{a}+\frac{1}{b}+\frac{1}{c}+\frac{1}{d} \geq 2.$$
But this is a consequence of the AM-HM inequality
$$\frac{1}{a}+\frac{1}{b}+\frac{1}{c}+\frac{1}{d} \geq \frac{16}{a+b+c+d} = 2.$$

Example 7.15. Find all nonnegative real numbers x, y, z, w satisfying simultaneously the following inequalities:

$$x + y + z \leq w, \quad x^2 + y^2 + z^2 \geq w, \quad x^3 + y^3 + z^3 \leq w.$$

Solution. Start by writing down the Cauchy-Schwarz inequality:

$$(x + y + z)(x^3 + y^3 + z^3) \geq (x^2 + y^2 + z^2)^2.$$

By hypothesis the left hand-side does not exceed w^2, while the right hand-side is greater than or equal to w^2. We deduce that we must have equality everywhere, including the Cauchy-Schwarz inequality. This happens if and only if there is c such that $x^3 = cx$, $y^3 = cy$, $z^3 = cz$. On the other hand, we must have

$$x + y + z = w, \quad x^2 + y^2 + z^2 = w, \quad x^3 + y^3 + z^3 = w.$$

The last relation can be written $c(x + y + z) = w$, which combined with the first relation yields $cw = w$. Thus either $c = 1$ or $w = 0$. If $w = 0$, then $x = y = z = 0$, which is a solution. Otherwise $x^3 = x$, $y^3 = y$ and $z^3 = z$, which means that $x, y, z \in \{0, 1\}$. Conversely, if $x, y, z \in \{0, 1\}$, then

$$x + y + z = x^2 + y^2 + z^2 = x^3 + y^3 + z^3,$$

hence we can take $w = x + y + z$. We conclude that the solutions of the problem are $(x, y, z, x + y + z)$, with $x, y, z \in \{0, 1\}$.

Example 7.16. Determine all real numbers a for which there are nonnegative real numbers $x_1, x_2, ..., x_5$ which satisfy

$$\sum_{k=1}^{5} kx_k = a, \quad \sum_{k=1}^{5} k^3 x_k = a^2, \quad \sum_{k=1}^{5} k^5 x_k = a^3.$$

Solution. Suppose that $a, x_1, ..., x_5$ satisfy the relations in the problem. Then

$$(x_1 + 2x_2 + ... + 5x_5)(x_1 + 2^5 x_2 + ... + 5^5 x_5) = (x_1 + 2^3 x_2 + ... + 5^3 x_5)^2,$$

since both terms are equal to a^4, by assumption. On the other hand, the Cauchy-Schwarz inequality yields

$$(x_1 + 2x_2 + ... + 5x_5)(x_1 + 2^5 x_2 + ... + 5^5 x_5)$$
$$\geq (\sqrt{x_1 \cdot x_1} + \sqrt{2x_2 \cdot 2^5 x_2} + ... + \sqrt{5x_5 \cdot 5^5 x_5})^2$$
$$= (x_1 + 2^3 x_2 + ... + 5^3 x_5)^2.$$

Since by assumption this is an equality, it follows that we must have equality in the Cauchy-Schwarz inequality. Hence there is a real number t such that $k^5 x_k = tkx_k$ for $1 \leq k \leq 5$. This can be written as

$$kx_k(t - k^4) = 0.$$

Since we can have $t - k^4 = 0$ for at most one value of k, it follows that at least four of the numbers $x_1, ..., x_5$ are 0. If they are all 0, then $a = 0$, which is a solution of the problem. If not, say $x_k \neq 0$. The equations become $kx_k = a$, $k^3 x_k = a^2$ and $k^5 x_k = a^3$. Thus $x_k = \frac{a}{k}$ and thus $k^2 a = a^2$ and $k^4 a = a^3$. Since we assumed that $a \neq 0$, we see that we must have $a = k^2$ and $x_k = k$. We conclude that the solutions of the problem are the numbers $0, 1^2, 2^2, ..., 5^2$.

Example 7.17. Prove that for all positive real numbers a, b, c we have

$$\frac{a^3}{b^2} + \frac{b^3}{c^2} + \frac{c^3}{a^2} \geq \frac{a^2}{b} + \frac{b^2}{c} + \frac{c^2}{a}.$$

Solution. Using the Cauchy-Schwarz inequality, we obtain

$$\left(\frac{a^2}{b} + \frac{b^2}{c} + \frac{c^2}{a}\right)^2 = \left(\sqrt{\frac{a^3}{b^2}} \cdot \sqrt{a} + \sqrt{\frac{b^3}{c^2}} \cdot \sqrt{b} + \sqrt{\frac{c^3}{a^2}} \cdot \sqrt{c}\right)^2$$

$$\leq \left(\frac{a^3}{b^2} + \frac{b^3}{c^2} + \frac{c^3}{a^2}\right) \cdot (a + b + c).$$

It suffices therefore to show that

$$a + b + c \leq \frac{a^2}{b} + \frac{b^2}{c} + \frac{c^2}{a},$$

which is again a consequence of the Cauchy-Schwarz inequality in the form of corollary 7.4

$$\frac{a^2}{b} + \frac{b^2}{c} + \frac{c^2}{a} \geq \frac{(a+b+c)^2}{b+c+a} = a + b + c.$$

Example 7.18. Let a, b, c be positive real numbers such that $a^2 + b^2 + c^2 = 3$. Prove that

$$(a + b + c)\left(\frac{a}{b} + \frac{b}{c} + \frac{c}{a}\right) \geq 9.$$

Solution. The Cauchy-Schwarz inequality in the form of corollary 7.4 yields

$$\frac{a}{b} + \frac{b}{c} + \frac{c}{a} \geq \frac{(a+b+c)^2}{ab+bc+ca}.$$

Hence it suffices to prove the inequality

$$(a+b+c)^3 \geq 9(ab+bc+ca).$$

On the other hand, the AM-GM inequality combined with the hypothesis give

$$3(ab+bc+ca)^2 = (a^2+b^2+c^2)(ab+bc+ca)^2 \leq \left(\frac{a^2+b^2+c^2+2(ab+bc+ca)}{3}\right)^3 = \left(\frac{(a+b+c)^2}{3}\right)^3.$$

Multiplying this inequality by 27 and taking square roots we obtain exactly the desired inequality

$$(a+b+c)^3 \geq 9(ab+bc+ca).$$

8 Making linear combinations

Many algebra problems given in math competitions look very random and do not have any apparent symmetry. For instance, we are given a set of relations between real numbers that are not homogeneous, or symmetric, and may involve more unknowns than given relations. In this case making linear combinations of those relations is a good start. The goal is to simplify those relations by adding/subtracting or performing more complicated linear combinations of them. Often, if you do these operations in the right way, you will find simpler expressions that can be completed to sums of squares or that can be nicely factored. Of course, the difficult part of the job is finding those linear combinations that simplify the problem. There is no recipe for this, and only enough training and experience will help from this point of view. We will give a long list of examples that will hopefully help in this process.

Example 8.1. Solve in real numbers the system of equations

$$\begin{cases} x - y^2 - z = \frac{1}{3} \\ y - z^2 - x = \frac{1}{6}. \end{cases}$$

Solution. Adding the two equations of the system yields

$$y - z^2 - y^2 - z = \frac{1}{3} + \frac{1}{6} = \frac{1}{2}.$$

Rewrite this by separating the variables

$$y^2 - y + z^2 + z + \frac{1}{2} = 0,$$

then complete the square to obtain

$$\left(y - \frac{1}{2}\right)^2 + \left(z + \frac{1}{2}\right)^2 = 0$$

and then $y = \frac{1}{2}$ and $z = -\frac{1}{2}$. Since $x = y^2 + z + \frac{1}{3}$, it follows that the system has a unique solution, given by $(x, y, z) = \left(\frac{1}{12}, \frac{1}{2}, -\frac{1}{2}\right)$.

Example 8.2. Find all triples (x, y, z) of real numbers satisfying the system of equations

$$\begin{cases} 2x - z^2 = -7 \\ 4y - x^2 = 7 \\ 6z - y^2 = 14. \end{cases}$$

Solution. One would be tempted to simply express x in terms of z from the first equation, then y in terms of x from the second equation, and replacing this in the third equation. That would end up in an eighth degree equation in z, with rather big coefficients. Thus this is certainly not the way to solve this problem.

We add up the equations and then separate the variables, arriving at
$$2x - x^2 + 4y - y^2 + 6z - z^2 = 14,$$
which can also be written as
$$x^2 - 2x + y^2 - 4y + z^2 - 6z + 14 = 0.$$
It is rather apparent how to complete the squares in the previous equation, obtaining
$$(x-1)^2 + (y-2)^2 + (z-3)^2 = 0$$
and the unique solution $(x, y, z) = (1, 2, 3)$ (we easily check that this is indeed a solution of the system).

Example 8.3. Find all real numbers x, y such that
$$\begin{cases} x^3 - 3xy^2 + 2y^3 = 2011 \\ 2x^3 - 3x^2y + y^3 = 2012. \end{cases}$$

Solution. Subtracting the second equation from the first we obtain
$$(y-x)^3 = -1, \quad \text{thus} \quad x = 1 + y.$$
Replacing this value of x in the first equation we obtain
$$1 + 3y + 3y^2 + y^3 - 3y^2 - 3y^3 + 2y^3 = 2011,$$
which simplifies rather miraculously to
$$3y = 2010, \quad \text{hence} \quad y = \frac{2010}{3} = 670.$$
Thus $x = 671$ and we the system has the unique solution $(671, 670)$. Note that instead of replacing $x = 1 + y$ in the first equation, we could have also added the equations, arriving at
$$x^3 - x^2y - xy^2 + y^3 = \frac{2011 + 2012}{3} = 1341.$$
The left hand-side factors as
$$x^2(x-y) - y^2(x-y) = (x-y)^2(x+y) = x+y,$$
since we have already seen that $x - y = 1$. Thus $x + y = 1341$ and $x - y = 1$, yielding the solution $(671, 670)$.

Example 8.4. Solve in real numbers the system of equations

$$\begin{cases} x^2 - yz + zx + xy = 6 \\ y^2 - zx + xy + yz = 19 \\ z^2 - xy + yz + zx = 30. \end{cases}$$

Solution. The system looks fairly complicated, but if we add the first two equations we obtain the much simpler relation

$$x^2 + 2xy + y^2 = 25,$$

that is $(x+y)^2 = 25$ and then $x + y = \pm 5$.

Similarly, adding the first and the third, then the second and the third equations, we obtain

$$x + z = \pm 6, \quad y + z = \pm 7.$$

Conversely, if these relations are satisfied, then (x, y, z) is a solution of the system. Hence we only need to solve each of the systems

$$\begin{cases} x + y = a \\ y + z = b \\ z + x = c, \end{cases}$$

with $a = \pm 5$, $b = \pm 7$ and $c = \pm 6$. In order to solve this system, we add up the equations to get

$$x + y + z = \frac{a+b+c}{2},$$

then we compare this relation with each of the equations of the system. We obtain

$$z = \frac{b+c-a}{2}, \quad x = \frac{c+a-b}{2}, \quad y = \frac{a+b-c}{2}.$$

Finally, taking all the previous values of a, b, c yields the solutions

$$(x, y, z) = (\pm 9, \mp 4, \mp 3), \quad (\pm 4, \mp 9, \pm 2), \quad (\pm 3, \pm 2, \mp 9), \quad (\pm 2, \pm 3, \pm 4).$$

Example 8.5. Find all triples (x, y, z) of real numbers such that

$$\begin{cases} (x+y)(x+y+z) = 72 \\ (y+z)(x+y+z) = 96 \\ (z+x)(x+y+z) = 120. \end{cases}$$

Solution. Adding these together gives
$$(2x + 2y + 2z)(x + y + z) = 288 \Leftrightarrow x + y + z = \pm 12.$$

Thus $x + y + z = 12k$, with $k \in \{-1, 1\}$. The system becomes
$$\begin{cases} x + y = 6k \\ y + z = 8k \\ z + x = 10k. \end{cases}$$

Since $x + y + z = 12k$, the first equation gives $12k - z = 6k$, thus $z = 6k$. Similarly, $y = 2k$ from the third equation and $x = 4k$ from the second one. We conclude that the solutions are $(x, y, z) = (4, 2, 6), (-4, -2, -6)$.

Example 8.6. Solve in real numbers
$$\begin{cases} 2x^2 + 3xy - 4y^2 = 1 \\ 3x^2 + 4xy - 5y^2 = 2. \end{cases}$$

Solution. Let us denote $y = tx$. Then the two equations can also be written as
$$\begin{cases} 2 + 3t - 4t^2 = \frac{1}{x^2} \\ 3 + 4t - 5t^2 = \frac{2}{x^2}. \end{cases}$$

This is quite easy to solve, by eliminating the variable x. We can do this by subtracting from the second equation twice the first equation. We obtain
$$3 + 4t - 5t^2 - 2(2 + 3t - 4t^2) = 0,$$

which can be simplified to $3t^2 - 2t - 1 = 0$, with the solutions $t = 1$ and $t = -\frac{1}{3}$. If $t = 1$, then coming back to the previous system gives $x^2 = 1$, thus $x = \pm 1$ and $x = y$. If $t = -\frac{1}{3}$, the same argument gives $x^2 = \frac{9}{5}$ and $x = \pm\frac{3}{\sqrt{5}}$, $y = -\frac{x}{3}$. We conclude that the solutions are

$$(x, y) = \left(\frac{-3}{\sqrt{5}}, \frac{1}{\sqrt{5}}\right), \quad \left(\frac{3}{\sqrt{5}}, \frac{-1}{\sqrt{5}}\right), \quad (1, 1), \quad (-1, -1)$$

Example 8.7. Find all positive real numbers x, y, z such that
$$\begin{cases} x + \frac{1}{y} = 4 \\ y + \frac{4}{z} = 3 \\ z + \frac{9}{x} = 5. \end{cases}$$

Solution. There is a smart, short, but hard to find solution, and a longer, but easy to find solution. Let us start with the short one: we simply add up the equations and rearrange terms to separate the variables. We obtain

$$x + \frac{9}{x} + y + \frac{1}{y} + z + \frac{4}{z} = 12.$$

Next, we complete the squares to get

$$\frac{(x-3)^2}{x} + \frac{(y-1)^2}{y} + \frac{(z-2)^2}{z} = 0,$$

which gives the only solution $(x, y, z) = (3, 1, 2)$ of the system.

Let us give now the less elegant solution, which however works in greater generality. We express $y = \frac{1}{4-x}$ from the first equation, then

$$z = \frac{4}{3-y} = \frac{4}{3 - \frac{1}{4-x}} = \frac{4(4-x)}{11 - 3x}.$$

Finally, replacing these values in the last equation yields

$$\frac{4(4-x)}{11-3x} + \frac{9}{x} = 5.$$

Clearing denominators gives the quadratic equation

$$16x - 4x^2 + 99 - 27x = 55x - 15x^2,$$

which simplifies to $11x^2 - 66x + 99 = 0$. Dividing by 11 we obtain $x^2 - 6x + 9 = 0$, or $(x-3)^2 = 0$. Thus $x = 3$ and coming back to the formulae giving y and z in terms of x, we obtain $y = 1$ and $z = 2$.

Example 8.8. Find all pairs (x, y) of real numbers satisfying

$$\begin{cases} x^2 - xy = 90 \\ y^2 - xy = -9. \end{cases}$$

Solution. Write the equations as $x(x - y) = 90$ and $y(y - x) = -9$. In particular, we must have $y \neq 0$. Dividing the first by the second, we obtain

$$\frac{x}{y} = \frac{x(x-y)}{-y(y-x)} = 10.$$

Thus $x = 10y$. Inserting this in the first equation yields $100y^2 - 10y^2 = 90$, thus $y^2 = 1$ and then $y = \pm 1$. Since $x = 10y$, we obtain the two solutions $(x, y) = (10, 1)$ and $(x, y) = (-10, -1)$ of the system.

Here is an alternative approach. We start by adding up the equations. This gives $(x - y)^2 = 81$, that is $x - y = \pm 9$. Since $x(x - y) = 90$, this shows that $x = \pm 10$ and we can easily find y in a similar way.

Example 8.9. Find all triples (x, y, z) of positive integers satisfying the system of equations
$$\begin{cases} x + yz = 2011 \\ xy + z = 2012. \end{cases}$$

Solution. Subtracting the two equations gets rid of the pretty large numbers involved. We obtain
$$xy + z - x - yz = 1$$
and the good news is that the left hand-side factors nicely as
$$x(y-1) - z(y-1) = (x-z)(y-1).$$

Thus $(x-z)(y-1) = 1$ and since we supposed that y is positive, it follows that we must have $y - 1 = 1$ and $x - z = 1$. That is, $y = 2$ and $x = z + 1$. Coming back to the first equation of the system yields
$$z + 1 + 2z = 2011$$
with the unique solution $z = 670$. Since $x = z + 1$, we conclude that the system has a unique solution, namely $(x, y, z) = (671, 2, 670)$.

Example 8.10. Find all triples (x, y, z) of positive real numbers such that
$$\begin{cases} \frac{xy}{x+y} = \frac{1}{5} \\ \frac{yz}{y+z} = \frac{1}{8} \\ \frac{zx}{z+x} = \frac{1}{9}. \end{cases}$$

Solution. Let us flip every equation in the system. Since
$$\frac{1}{\frac{xy}{x+y}} = \frac{1}{x} + \frac{1}{y},$$
we obtain the equivalent system
$$\begin{cases} \frac{1}{x} + \frac{1}{y} = 5 \\ \frac{1}{z} + \frac{1}{y} = 8 \\ \frac{1}{x} + \frac{1}{z} = 9 \end{cases}$$

The good news is that this is a linear (and very simple) system in the unknowns $\frac{1}{x}, \frac{1}{y}$ and $\frac{1}{z}$. Adding up the equations and dividing by 2 gives
$$\frac{1}{x} + \frac{1}{y} + \frac{1}{z} = 11.$$

Combining this with the equations of the system, we deduce that

$$\frac{1}{x} = 3, \quad \frac{1}{y} = 2, \quad \frac{1}{z} = 6,$$

hence the unique solution

$$x = \frac{1}{3}, \quad y = \frac{1}{2}, \quad z = \frac{1}{6}.$$

Example 8.11. If a, b, c are real numbers such that

$$(a-b)(a+b-c) = 3 \quad \text{and} \quad (b-c)(b+c-a) = 5,$$

find $(c-a)(c+a-b)$.

Solution. We have two equations and three variables, so there must be some subtlety here! We guess that there should be some hidden relation among the numbers $(a-b)(a+b-c)$, $(b-c)(b+c-a)$ and $(c-a)(c+a-b)$. Well, this is not so hard to see when we expand:

$$(a-b)(a+b-c) = (a-b)(a+b) - (a-b)c = a^2 - b^2 - (ac - bc),$$
$$(b-c)(b+c-a) = (b-c)(b+c) - (b-c)a = b^2 - c^2 - (ab - ac),$$
$$(c-a)(c+a-b) = (c-a)(c+a) - (c-a)b = c^2 - a^2 - (cb - ab).$$

Adding these relations yields

$$(a-b)(a+b-c) + (b-c)(b+c-a) + (c-a)(c+a-b) = 0.$$

Taking into account the hypothesis, we deduce that

$$(c-a)(c+a-b) = -(3+5) = -8.$$

Example 8.12. Find all triples (x, y, z) of real numbers such that

$$\begin{cases} x^2 = 4(y-1) \\ y^2 = 4(z-1) \\ z^2 = 4(x-1). \end{cases}$$

Solution. The easiest way is to simply add up the equations and separate the variables. We obtain

$$x^2 - 4x + y^2 - 4y + z^2 - 4z + 12 = 0.$$

We complete the squares for each variable and obtain

$$(x-2)^2 + (y-2)^2 + (z-2)^2 = 0.$$

Thus we must have $x = y = z = 2$ and we easily check that this is indeed a solution of the system, thus the unique solution.

Example 8.13. Solve in real numbers the system of equations
$$\begin{cases} x + y^2 = \frac{1}{12} \\ y - 3z^2 = \frac{1}{4} \\ z - \frac{3}{2}x^2 = \frac{1}{3}. \end{cases}$$

Solution. We multiply the first equation by 2, the second and third equations by -2, obtaining the system
$$\begin{cases} 2x + 2y^2 = \frac{1}{6} \\ -2y + 6z^2 = -\frac{1}{2} \\ -2z + 3x^2 = -\frac{2}{3}. \end{cases}$$

Adding up these equations, and separating the variables, we deduce that
$$3x^2 + 2x + 2y^2 - 2y + 6z^2 - 2z = -1.$$

Next, we complete the squares and obtain
$$3\left(x + \frac{1}{3}\right)^2 + 2\left(y - \frac{1}{2}\right)^2 + 6\left(z - \frac{1}{6}\right)^2 = 0.$$

This shows that any solution must satisfy
$$(x, y, z) = \left(-\frac{1}{3}, \frac{1}{2}, \frac{1}{6}\right).$$

However, if we plug in these values in the original equations of the system, we see that the equations are not satisfied, hence the system has no real solution.

Example 8.14. Solve in real numbers the system of equations
$$\begin{cases} x^2 + 6xy + 2y^2 = 1 \\ y^2 + 6yz + 2z^2 = 2 \\ z^2 + 6zx + 2x^2 = -3. \end{cases}$$

Solution. The key step is adding the equations to obtain
$$3(x^2 + y^2 + z^2) + 6(xy + yz + zx) = 0,$$

which after division by 3 becomes simply $(x+y+z)^2 = 0$. Thus $x+y+z = 0$, which allows us to replace $x = -y - z$ in the first equation
$$(y+z)^2 - 6y(y+z) + 2y^2 = 1.$$

Making linear combinations

This simplifies after expansion and rearrangement to
$$z^2 - 4yz - 3y^2 = 1.$$

We combine this with the second equation
$$y^2 + 6yz + 2z^2 = 2.$$

Subtracting from this second equation twice the relation previously established, we obtain
$$y^2 + 6yz + 2z^2 - 2(z^2 - 4yz - 3y^2) = 0,$$

which simplifies to $7y^2 + 14yz = 0$ and then $y = 0$ or $y = -2z$.

Suppose that $y = 0$, then the equations become $z^2 = 1$ and $x^2 = 1$, with $x = -z$. We obtain the solutions $(x, y, z) = \{(1, 0, -1), (-1, 0, 1)\}$. Suppose finally that $y = -2z$. Then the second equation of the system becomes $4z^2 - 12z^2 + 2z^2 = 2$, or $-6z^2 = 2$, which has no real solutions. Thus the solutions in real numbers are $(x, y, z) = \{(1, 0, -1), (-1, 0, 1)\}$.

Let us come back a little bit to the first step, which is real the crucial one. The natural question is: why would we add the equations? Let us set $y = tx$ and $z = wx$. Then the system becomes
$$\begin{cases} 1 + 6t + 2t^2 = \frac{1}{x^2} \\ t^2 + 6tw + 2w^2 = \frac{2}{x^2} \\ w^2 + 6w + 2 = -\frac{3}{x^2}. \end{cases}$$

Now we try to eliminate the variable x, and the easiest way is to add up the equations.

Example 8.15. Solve in real numbers the system of equations
$$\begin{cases} \frac{xyz}{x+y} = -2 \\ \frac{xyz}{y+z} = 3 \\ \frac{xyz}{z+x} = 6. \end{cases}$$

Solution. Note that $xyz \neq 0$. We rewrite this as a system
$$\begin{cases} x + y = -\frac{xyz}{2} \\ y + z = \frac{xyz}{3} \\ z + x = \frac{xyz}{6}. \end{cases}$$

Adding these relations yields
$$2(x + y + z) = 0,$$

hence
$$x+y=-z, \quad y+z=-x, \quad z+x=-y.$$
Going back to the previous system, we obtain
$$\begin{cases} z = \frac{xyz}{2} \\ x = -\frac{xyz}{3} \\ y = -\frac{xyz}{6}. \end{cases}$$
Multiplying these relations, we end up with
$$xyz = \frac{(xyz)^3}{36}.$$
Since $xyz \neq 0$, the previous equation is equivalent to $xyz = \pm 6$. Replacing these values in the previous system finally yields the solutions $(x, y, z) = (-2, -1, 3)$ and $(x, y, z) = (2, 1, -3)$.

Another approach is to start by inverting the equations to get linear equations in the unknowns $\frac{1}{xy}$, $\frac{1}{yz}$, and $\frac{1}{xz}$. Solving gives $\frac{1}{xy} = \frac{1}{2}$, $\frac{1}{yz} = -\frac{1}{3}$, and $\frac{1}{xz} = -\frac{1}{6}$, or $xy = 2$, $yz = -3$ and $zx = -6$. From here it is not difficult to obtain the two solutions of the system.

Example 8.16. Find all real numbers x, y, z such that
$$\begin{cases} x^2 - 3 = (y-z)^2 \\ y^2 - 5 = (z-x)^2 \\ z^2 - 15 = (x-y)^2 \end{cases}$$

Solution. This is rather tricky. We rewrite the system as
$$\begin{cases} x^2 - (y-z)^2 = 3 \\ y^2 - (z-x)^2 = 5 \\ z^2 - (x-y)^2 = 15 \end{cases}$$
and exploit the difference of squares formula to obtain the equivalent system
$$\begin{cases} (x-y+z)(x+y-z) = 3 \\ (x+y-z)(-x+y+z) = 5 \\ (x-y+z)(-x+y+z) = 15 \end{cases}$$

Let us denote
$$a = y+z-x, \quad b = z+x-y, \quad c = x+y-z.$$

Then
$$bc = 3, \quad ca = 5, \quad ab = 15.$$

Taking the product of these relations yields $(abc)^2 = 15^2$, hence $abc = \pm 15$. Combining this with the previous relations, we conclude that $(a, b, c) = (5, 3, 1)$ or $(a, b, c) = (-5, -3, -1)$.

Next, we have
$$y + z - x = a, \quad z + x - y = b, \quad x + y - z = c,$$

that is x, y, z are the solutions of a linear system. We can solve it quite easily by adding the equations, which gives $x + y + z = a + b + c$. Thus
$$a = y + z - x = x + y + z - 2x = a + b + c - 2x,$$

giving $x = \frac{b+c}{2}$. Similarly, we obtain $y = \frac{c+a}{2}$ and $z = \frac{a+b}{2}$. Since $(a, b, c) = (5, 3, 1)$ or $(a, b, c) = (-5, -3, -1)$, we finally conclude that the solutions are $(x, y, z) = (2, 3, 4)$ and $(x, y, z) = (-2, -3, -4)$.

Example 8.17. Solve in real numbers the system of equations
$$\begin{cases} x(y^2 + zx) = 111 \\ y(z^2 + xy) = 155 \\ z(x^2 + yz) = 116. \end{cases}$$

Solution. Expand each term in the left hand-side of the equations to obtain the system
$$\begin{cases} xy^2 + zx^2 = 111 \\ yz^2 + xy^2 = 155 \\ zx^2 + yz^2 = 116. \end{cases}$$

This is a linear system in the variables xy^2, yz^2 and zx^2, hence it can easily be solved. The quickest way is to add up the equations to get
$$xy^2 + yz^2 + zx^2 = 191$$

and then comparing this new relation with each of the equations of the previous system. This gives
$$xy^2 = 75, \quad yz^2 = 80, \quad zx^2 = 36,$$

which is quite easy to solve: by multiplying the previous relations, we obtain
$$x^3 y^3 z^3 = 75 \cdot 80 \cdot 36.$$

We try to factor the right hand-side in order to easily extract the cube root. Write
$$75 \cdot 80 \cdot 36 = 3 \cdot 5^2 \cdot 2^4 \cdot 5 \cdot 2^2 \cdot 3^2 = 2^6 \cdot 3^3 \cdot 5^3.$$
We deduce that
$$xyz = 2^2 \cdot 3 \cdot 5 = 60.$$
Next, rewrite the relation $xy^2 = 75$ as follows
$$75 = xy^2 = xy \cdot y = \frac{60}{z} \cdot y,$$
giving $\frac{y}{z} = \frac{5}{4}$. We do the same with the other two relations and we obtain
$$\frac{z}{x} = \frac{4}{3}, \quad \frac{x}{y} = \frac{3}{5}, \quad \frac{y}{z} = \frac{5}{4}.$$
This allows us to express y and z in terms of x, as follows
$$z = \frac{4x}{3}, \quad y = \frac{5x}{3}.$$
Finally, replacing this in $xyz = 60$ yields $x = 3$, then $z = 4$ and $y = 5$. Thus the system has a unique solution, $(x, y, z) = (3, 5, 4)$.

Example 8.18. Given $a, b, c > 0$, find all positive real numbers x, y, z such that
$$\begin{cases} ax - by + \frac{1}{xy} = c \\ bz - cx + \frac{1}{zx} = a \\ cy - az + \frac{1}{yz} = b. \end{cases}$$

Solution. We will consider this as a system in the unknowns a, b, c, the reason being that it is then a linear system. We can easily solve it: we replace $b = cy - az + \frac{1}{yz}$ in the first two equations of the system and obtain
$$a = \frac{1}{xz}, \quad b = \frac{1}{yz}, \quad c = \frac{1}{xy}.$$
It is now easy to solve the system: multiplying the equations gives
$$abc = \frac{1}{(xyz)^2}, \quad \text{hence} \quad xyz = \frac{1}{\sqrt{abc}},$$
thus
$$a = \frac{y}{xyz} = y\sqrt{abc}, \quad \text{thus} \quad y = \sqrt{\frac{a}{bc}}.$$
Similarly, the obtain the unique solution
$$x = \sqrt{\frac{b}{ac}}, \quad y = \sqrt{\frac{a}{bc}}, \quad z = \sqrt{\frac{c}{ab}}.$$

9 Fixed points and monotonicity

There is a class of equations and systems of equations which are rather hard to solve by conventional means (i.e. simple algebraic manipulations), but which have very elegant solutions using the monotonicity of certain functions. In this section we will establish a general method of solving such problems and give some examples.

Let us recall a few definitions first. Let X be a subset of $\mathbf{R}$ and let $f : X \to \mathbf{R}$ be a real-valued map on X. We say that f is
- increasing if $f(x) < f(y)$ for all $x < y \in X$.
- nondecreasing if $f(x) \leq f(y)$ for all $x < y \in X$.
- decreasing if $f(x) > f(y)$ for all $x < y \in X$.
- injective when $f(x) \neq f(y)$ whenever $x \neq y \in X$.

The fundamental property of injective maps is that an equation $f(x) = a$ (with a a given real number) has at most one solution. Indeed, if x_1, x_2 are solutions, then $f(x_1) = f(x_2)$ (since both are equal to a), hence by injectivity $x_1 = x_2$. Thus if by some way we manage to find a solution of the equation, then we are done: this will be the unique solution of the equation.

It is quite hard to establish the injectivity of a map in general, but very often this map is increasing or decreasing, and then injectivity is automatic (this is a direct consequence of the definitions). In order to easily recognize increasing functions, it is important to have a good supply of functions we already know are increasing or decreasing. Here are some examples:
- The function $x \to x^{2n+1}$ is increasing on $\mathbf{R}$ for all $n \geq 0$. The function $x \to x^{2n}$ is increasing on $[0, \infty)$ and decreasing on $(-\infty, 0]$.
- If $n > 0$, then the function $x \to \frac{1}{x^n}$ is decreasing on $(0, \infty)$.
- If n is odd, then the function $x \to \sqrt[n]{x}$ is increasing on $\mathbf{R}$, and if n is even, then $x \to \sqrt[n]{x}$ is increasing on $[0, \infty)$.
- The sum of two increasing (resp. decreasing) functions is increasing (resp. decreasing).
- The composition of two increasing functions is increasing, and so is the composition of two decreasing functions. Indeed, if f, g are increasing (resp. decreasing) and if $x < y$, then $f(x) < f(y)$ (resp. $f(x) > f(y)$), hence $g(f(x)) < g(f(y))$ (resp. $g(f(x)) < g(f(y))$), thus $g \circ f$ is increasing. With the same argument we can prove that the composition of a decreasing and an increasing map is decreasing.

Here is now a fundamental result. For simplicity, we denote $f^{[n]} = f \circ f \circ ... \circ f$ the nth iterate of f.

Theorem 9.1. *Let $f : X \to X$ be a map on some nonempty subset X of $\mathbf{R}$, and let $x \in X$ satisfy $f^{[n]}(x) = x$ for some $n \geq 1$.*
  a) If f is increasing, then $f(x) = x$.

b) If f is decreasing, then $f(f(x)) = x$. Moreover, if n is odd, then $f(x) = x$.

Proof. a) Let us suppose that $f(x) < x$. Applying f and using the fact that f is increasing, we obtain $f(f(x)) < f(x)$. Applying f again yields

$$f^{[3]}(x) < f^{[2]}(x) < f(x) < x.$$

Continuing like this we obtain

$$x = f^{[n]}(x) < f^{[n-1]}(x) < \ldots < f(x) < x,$$

a contradiction. Similarly, if $f(x) > x$, then

$$x = f^{[n]}(x) > f^{[n-1]}(x) > \ldots > f(x) > x,$$

again a contradiction. Thus we must have $f(x) = x$, proving part a) of the theorem.

b) Note that

$$f^{[2n]}(x) = f^{[n]}(f^{[n]}(x)) = f^{[n]}(x) = x.$$

On the other hand, if $g = f \circ f$, then $f^{[2n]}(x) = g^{[n]}(x)$. Thus $g^{[n]}(x) = x$. But since f is decreasing, $g = f \circ f$ is increasing, hence we may apply part a) of the theorem to g and obtain $g(x) = x$, that is $f(f(x)) = x$. Finally, if n is odd, then $f^{[n-1]}(x) = x$ (since $f(f(x)) = x$ and $n - 1$ is even), hence

$$x = f^{[n]}(x) = f(f^{[n-1]}(x)) = f(x)$$

and we are done. □

When solving systems of equations, the following corollary is very useful:

Corollary 9.2. *Let $f : X \to X$ be an increasing function. Then the solutions **in** X of the system*

$$\begin{cases} f(x_1) = x_2 \\ f(x_2) = x_3 \\ \ldots \\ f(x_n) = x_1. \end{cases}$$

are given by $(t, t, \ldots, t)$, where t is a solution of the equation $f(t) = t$.

Be very careful of the fact that this only gives the solutions in X. In practice, we try to first find a largest possible X so that f maps X to X and is increasing on X, and then try to prove that all solutions of the system actually lie in X. If we manage to do all this, then the resolution of the system comes down to solving the equation $f(t) = t$ in X, as the corollary shows.

Proof. If $(x_1, ..., x_n)$ is a solution of the system, then

$$x_1 = f(x_n) = f(f(x_{n-1})) = f^{[3]}(x_{n-2}) = ... = f^{[n]}(x_1),$$

thus $f^{[n]}(x_1) = x_1$. Applying the previous theorem yields $f(x_1) = x_1$, then $x_2 = f(x_1) = x_1$,..., $x_n = x_1$, and the corollary follows. □

Note that the previous corollary also applies to the case when n (the number of unknowns) is odd and f is decreasing on X (as follows from part b) of the previous theorem).

Example 9.1. Solve in real numbers the system

$$\begin{cases} 3a = (b+c+d)^3 \\ 3b = (c+d+a)^3 \\ 3c = (d+a+b)^3 \\ 3d = (a+b+c)^3. \end{cases}$$

Solution. Let us write the system in the equivalent form

$$\begin{cases} \sqrt[3]{3a} = b+c+d \\ \sqrt[3]{3b} = c+d+a \\ \sqrt[3]{3c} = d+a+b \\ \sqrt[3]{3d} = a+b+c. \end{cases}$$

We deduce that

$$a + \sqrt[3]{3a} = b + \sqrt[3]{3b} = c + \sqrt[3]{3c} = d + \sqrt[3]{3d} = a+b+c+d.$$

The key point is that the map $x \mapsto x + \sqrt[3]{3x}$ is increasing, in particular injective, since it is the sum of two increasing maps. Thus the previous equalities yield $a = b = c = d$ and coming back to the system we have $3a = (3a)^3 = 27a^3$. Thus $a = 0$ or $a = \pm\frac{1}{3}$. This gives us the three solutions of the system.

Example 9.2. Find all triples (x, y, z) of real numbers greater than 1 that satisfy the system of equations

$$\begin{cases} \frac{x}{x-1} = \frac{4}{y} \\ \frac{y}{y-1} = \frac{4}{z} \\ \frac{z}{z-1} = \frac{4}{x} \end{cases}$$

Solution. Let $f(x) = \frac{4(x-1)}{x}$. Since

$$f(x) = 4 - \frac{4}{x}$$

and since $x \to -\frac{4}{x}$ is increasing on $(0, \infty)$, we deduce that $f(x)$ is an increasing function of x for $x \geq 1$. Using corollary 9.2, we obtain $x = y = z$ and $f(x) = x$. Now, the equation $f(x) = x$ is very easy to solve, since after multiplication by x it becomes $(x-2)^2 = 0$. Thus $x = 2$ and then $y = z = 2$. Hence the system has a unique solution, namely $(2, 2, 2)$.

Here is a trickier solution. Flipping the equations yields

$$\begin{cases} \frac{x-1}{x} = \frac{y}{4} \\ \frac{y-1}{y} = \frac{z}{4} \\ \frac{z-1}{z} = \frac{x}{4} \end{cases}$$

or equivalently

$$\begin{cases} 1 - \frac{1}{x} = \frac{y}{4} \\ 1 - \frac{1}{y} = \frac{z}{4} \\ 1 - \frac{1}{z} = \frac{x}{4} \end{cases}$$

Let us add up the equations and rearrange terms so that to separate the variables. We obtain

$$\frac{x}{4} + \frac{1}{x} + \frac{y}{4} + \frac{1}{y} + \frac{z}{4} + \frac{1}{z} = 3.$$

Completing the square finally yields the equation

$$\frac{(x-2)^2}{2x} + \frac{(y-2)^2}{2y} + \frac{(z-2)^2}{2z} = 0,$$

which forces $x = y = z = 2$. It is easy to check that this is indeed a solution of the system, hence it is the unique solution.

Example 9.3. Solve in real numbers the system of equations

$$\begin{cases} x^4 = 4y - 3 \\ y^4 = 4z - 3 \\ z^4 = 4x - 3. \end{cases}$$

Solution. The first equation shows that $y = \frac{x^4+3}{4} > 0$. Similarly, we obtain $x > 0$ and $z > 0$. Since $f(t) = \frac{t^4+3}{4}$ is an increasing function of t for $t > 0$, it follows from corollary 9.2 that $x = y = z$ and this common value is a solution of the equation $f(t) = t$. Now, let us solve the equation $f(t) = t$, or equivalently $t^4 + 3 = 4t$. Using the AM-GM inequality, we have

$$t^4 + 3 = t^4 + 1 + 1 + 1 \geq 4t,$$

with equality if and only if $t = 1$. Hence $t = 1$ is the unique solution of the equation and the system has the unique solution $(x, y, z) = (1, 1, 1)$.

Here is an alternative way of solving the equation $t^4 + 3 = 4t$. An obvious solution is $t = 1$, which means that we can factor $t - 1$. Let's us write

$$t^4 - 4t + 3 = t^4 - t - 3(t-1) = t(t^3 - 1) - 3(t-1) =$$
$$t(t-1)(t^2 + t + 1) - 3(t-1) = (t-1)(t^3 + t^2 + t - 3).$$

The good news is that $t^3 + t^2 + t - 3$ also vanishes at $t = 1$, thus we can still factor

$$t^3 + t^2 + t - 3 = t^3 - t^2 + 2t^2 - 2t + 3t - 3 = (t-1)(t^2 + 2t + 3).$$

All in all, we have established that

$$t^4 - 4t + 3 = (t-1)^2(t^2 + 2t + 3).$$

Note that $t^2 + 2t + 3 = (t+1)^2 + 2 > 0$ for all t, thus $t^4 - 4t + 3 = 0$ is equivalent to $t = 1$.

Example 9.4. Find all triples (x, y, z) of real numbers satisfying the system of equations

$$\begin{cases} x^2 = 4(y-1) \\ y^2 = 4(z-1) \\ z^2 = 4(x-1). \end{cases}$$

Solution. The easiest way is to simply add up the equations and separate the variables. We obtain

$$x^2 - 4x + y^2 - 4y + z^2 - 4z + 12 = 0.$$

We complete the squares for each variable and obtain

$$(x-2)^2 + (y-2)^2 + (z-2)^2 = 0.$$

Thus we must have $x = y = z = 2$ and we easily check that this is indeed a solution of the system, thus the unique solution.

Here is an alternative approach. Write the system in the equivalent form

$$\begin{cases} y = 1 + \frac{x^2}{4} \\ z = 1 + \frac{y^2}{4} \\ x = 1 + \frac{z^2}{4}. \end{cases}$$

This clearly shows that $x, y, z > 0$ and that $y = f(x)$, $z = f(y)$, $x = f(z)$, where $f(x) = 1 + \frac{x^2}{4}$. Now, f is clearly increasing on $(0, \infty)$ and corollary 9.2 allows us to conclude that $x = y = z$ are solutions of the equation $f(x) = x$. This quadratic equation is equivalent to $(x-2)^2 = 0$, thus has a unique solution $x = 2$. Hence $x = y = z = 2$ is the unique solution of the system.

Example 9.5. Solve the equation
$$\sqrt{x+3} + \sqrt[3]{x+2} = 5.$$

Solution. Trying to take the square or cube of the given relation (or an equivalent one, obtained by taking one of the roots to the right hand-side) results in a complicated algebraic expression. The key point is that the left hand-side is an increasing, in particular injective map on $[-2, \infty)$. Indeed, each of the maps $x \mapsto \sqrt{x+3}$ and $x \to \sqrt[3]{x+2}$ is increasing, hence so is their sum. This argument already shows that the equation has at most one solution, so if we are lucky to find one, then we are done. Now, it is a simple matter of luck to find (or not) a solution. We can check that $x = 6$ is a solution (a good way to look for a solution in this case is to find an integer x for which $x+3$ is a square and $x+2$ is a cube; trial and error eventually leads to $x = 6$), hence it is the unique solution of the equation.

Example 9.6. Solve the system
$$\begin{cases} \sqrt{x} - \frac{1}{y} = \frac{7}{4} \\ \sqrt{y} - \frac{1}{z} = \frac{7}{4} \\ \sqrt{z} - \frac{1}{x} = \frac{7}{4}. \end{cases}$$

Solution. Write the system as $x = f(y)$, $y = f(z)$ and $z = f(x)$, where
$$f(t) = \left(\frac{7}{4} + \frac{1}{t}\right)^2.$$

Note that f is decreasing on $(0, \infty)$, since $t_1 < t_2$ implies $\frac{1}{t_1} > \frac{1}{t_2}$ and then $f(t_1) > f(t_2)$. The previous equations show that x, y, z are solutions of the equation $f(f(f(t))) = t$. By theorem 9.1, these are the same as the solutions of the equation $f(t) = t$. We conclude that $x = y = z$ are solutions of the equation
$$t = \left(\frac{7}{4} + \frac{1}{t}\right)^2.$$

This equation has the apparent solution $t = 4$, and no other positive solution, since the map $t \mapsto t - f(t)$ is increasing, thus injective. Hence the unique solution of the system is $(x, y, z) = (4, 4, 4)$.

10 The floor function

The floor function $\lfloor \cdot \rfloor : \mathbf{R} \to \mathbf{Z}$ is quite easy to define, but rather hard to study. By definition, $\lfloor x \rfloor$ is the greatest integer which does not exceed x. We also call $\lfloor x \rfloor$ the integer part of x. The fundamental property of $\lfloor x \rfloor$ is that it is an integer and
$$\lfloor x \rfloor \leq x < \lfloor x \rfloor + 1.$$

Note that this does not necessarily imply that $\lfloor x \rfloor$ is the integer closest to x. Indeed, for $x = \frac{3}{5}$ the integer closest to x is 1, but the integer part of x is 0. Actually, the integer closest to x is $\lfloor x + \frac{1}{2} \rfloor$. In practice, we often use the following characterization of the floor function: if n is an integer and x is a real number, then $\lfloor x \rfloor = n$ if and only if $x \in [n, n+1)$. It is therefore clear that the floor function is surjective and constant on each interval of the form $[n, n+1)$, with $n \in \mathbf{Z}$.

Another important map is the fractional part $\{\cdot\} : \mathbf{R} \to [0, 1)$. By definition, we have
$$\{x\} = x - \lfloor x \rfloor$$
for all x.

Let us give some crucial properties of the integer and fractional part maps, which will be constantly used throughout the examples:

Theorem 10.1. *a) We have* $\lfloor x + n \rfloor = \lfloor x \rfloor + n$ *for all **integers** n and all real numbers x.*

b) For all real numbers x, y
$$\lfloor x \rfloor + \lfloor y \rfloor \leq \lfloor x + y \rfloor \leq \lfloor x \rfloor + \lfloor y \rfloor + 1.$$

c) The map $\{\cdot\}$ is 1-periodic, i.e. $\{x + n\} = \{x\}$ for all real numbers x and all integers n.

d) The map $\{\cdot\}$ is sub-additive, i.e. $\{x+y\} \leq \{x\}+\{y\}$ for all real numbers x and y.

Proof. a) Suppose that $\lfloor x \rfloor = k$, so that $x \in [k, k+1)$. Then $x + n \in [n+k, n+k+1)$ and so
$$\lfloor x + n \rfloor = n + k = n + \lfloor x \rfloor.$$

b) Using part a), we can write
$$\lfloor x + y \rfloor = \lfloor \lfloor x \rfloor + \{x\} + \lfloor y \rfloor + \{y\} \rfloor = \lfloor x \rfloor + \lfloor y \rfloor + \lfloor \{x\} + \{y\} \rfloor.$$

Since $0 \leq \{x\} + \{y\} < 2$, it follows that $\lfloor \{x\} + \{y\} \rfloor \in \{0, 1\}$ and the result follows.

c) This follows from the definition of the fractional part and part a).

d) Written as
$$x - \lfloor x \rfloor + y - \lfloor y \rfloor \leq x + y - \lfloor x+y \rfloor,$$
this is easily seen to be equivalent equivalent to the lower bound in b). □

Another useful result concerning the floor function is the following

Theorem 10.2. *(Hermite's identity)* For all real numbers x and all positive integers n
$$\lfloor x \rfloor + \left\lfloor x + \frac{1}{n} \right\rfloor + ... + \left\lfloor x + \frac{n-1}{n} \right\rfloor = \lfloor nx \rfloor.$$

Proof. Using part a) of the previous theorem, we obtain the equivalent equality
$$n\lfloor x \rfloor + \left\lfloor \{x\} + \frac{1}{n} \right\rfloor + ... + \left\lfloor \{x\} + \frac{n-1}{n} \right\rfloor = n\lfloor x \rfloor + \lfloor n\{x\} \rfloor.$$

Let $k = \lfloor n\{x\} \rfloor$, so that $\{x\} \in [\frac{k}{n}, \frac{k+1}{n})$. Since $\{x\} < 1$, we must have $k < n$. If $j \in \{1, ..., n-1\}$, then $\lfloor \{x\} + \frac{j}{n} \rfloor \in \{0, 1\}$ and it is equal to 1 if and only if $\{x\} \geq 1 - \frac{j}{n}$. This happens if and only if $\frac{k}{n} \geq \frac{n-j}{n}$, that is $j \geq n - k$. We deduce that
$$\left\lfloor \{x\} + \frac{1}{n} \right\rfloor + ... + \left\lfloor \{x\} + \frac{n-1}{n} \right\rfloor = \sum_{j=n-k}^{n-1} 1 = k = \lfloor n\{x\} \rfloor$$
and the result follows. □

We will now give a long list of examples to see how these ideas can be used in practice.

Example 10.1. Prove that for all real numbers x and all positive integers n we have
$$\left\lfloor \frac{x}{n} \right\rfloor = \left\lfloor \frac{\lfloor x \rfloor}{n} \right\rfloor.$$

Solution. Let $k = \lfloor \frac{x}{n} \rfloor$, so that $\frac{x}{n} \in [k, k+1)$. Hence $x \in [nk, n(k+1))$. Since the floor function is nondecreasing and $n(k+1)$ is an integer, we obtain $\lfloor x \rfloor \in [nk, n(k+1))$ and so $\frac{\lfloor x \rfloor}{n} \in [k, k+1)$. We conclude that $\lfloor \frac{\lfloor x \rfloor}{n} \rfloor = k$, which is exactly what we wanted.

Example 10.2. Solve the equation
$$2x\{x\} = \lfloor x \rfloor + \frac{1}{2}.$$

Solution. Since $\lfloor x \rfloor + \{x\} = x$, we can write the equation as

$$2x\{x\} = x - \{x\} + \frac{1}{2},$$

then as

$$2x\{x\} - x + \{x\} - \frac{1}{2} = 0$$

and finally

$$(2x+1)\left(\{x\} - \frac{1}{2}\right) = 0.$$

Hence either $x = -\frac{1}{2}$ or $\{x\} = \frac{1}{2}$, which is to say $x = k + \frac{1}{2}$, with $k \in \mathbf{Z}$.

Example 10.3. Solve the equation

$$\lfloor x \rfloor^2 + 4\{x\}^2 = 4x - 5.$$

Solution. Since $x = \lfloor x \rfloor + \{x\}$, the equation can be written as

$$(\lfloor x \rfloor - 2)^2 + (2\{x\} - 1)^2 = 0.$$

This is equivalent to $\lfloor x \rfloor = 2$ and $\{x\} = \frac{1}{2}$, which yields the unique solution

$$x = \lfloor x \rfloor + \{x\} = 2 + \frac{1}{2} = \frac{5}{2}.$$

Example 10.4. Prove that for all real numbers x we have

$$\lfloor \sqrt[3]{x} \rfloor = \lfloor \sqrt[3]{\lfloor x \rfloor} \rfloor.$$

Solution. Let $n = \lfloor \sqrt[3]{x} \rfloor$, so that $\sqrt[3]{x} \in [n, n+1)$ and $x \in [n^3, (n+1)^3)$. Since the floor function is nondecreasing and $(n+1)^3$ is an integer, we obtain $\lfloor x \rfloor \geq n^3$ and $\lfloor x \rfloor < (n+1)^3$. Thus $\sqrt[3]{\lfloor x \rfloor} \in [n, n+1)$ and $\lfloor \sqrt[3]{\lfloor x \rfloor} \rfloor = n$. The result follows.

Example 10.5. Solve the equation $x^3 - \lfloor x \rfloor = 7$.

Solution. Let us denote $n = \lfloor x \rfloor$. Then the equation becomes $x^3 = n+7$, thus $x = \sqrt[3]{n+7}$. Next, the equality $n = \lfloor x \rfloor$ is equivalent to $n \leq x < n+1$. Thus $n^3 \leq n+7 < (n+1)^3$. Writing this as $n(n^2-1) \leq 7$ and $(n+1)((n+1)^2-1) > 6$, we see that we must have $n+1 \geq 3$ (from the second inequality) and $n \leq 2$ (from the first one). Thus $n = 2$ is the unique solution of the inequality and correspondingly $x = \sqrt[3]{9}$ is the unique solution of the equation.

Example 10.6. Solve the equation

$$x^2 + \lfloor x \rfloor^2 + \{x\}^2 = 10.$$

Solution. Let $n = \lfloor x \rfloor$, so that $x \in [n, n+1)$. The equation becomes

$$x^2 + n^2 + (x-n)^2 = 10, \quad \text{or} \quad x^2 - nx + n^2 - 5 = 0.$$

Since the discriminant of this quadratic equation is nonnegative, we must have $n^2 \geq 4(n^2 - 5)$, hence $3n^2 \leq 20$. This forces $n^2 \leq 6$ and so $n \in \{-2, -1, 0, 1, 2\}$. Next, solving the quadratic equation above yields

$$x = \frac{n + \sqrt{20 - 3n^2}}{2} \quad \text{or} \quad x = \frac{n - \sqrt{20 - 3n^2}}{2}.$$

Let us see when we have

$$n \leq \frac{n + \sqrt{20 - 3n^2}}{2} < n+1.$$

This is equivalent to

$$n \leq \sqrt{20 - 3n^2} < n + 2.$$

Since $n^2 \leq 4$, we have $\sqrt{20 - 3n^2} \geq \sqrt{8} = 2\sqrt{2}$ and so we must have $n + 2 > 2\sqrt{2}$, thus $n \geq 1$. It is easy to see that $n = 1$ is not a solution and $n = 2$ yields the solution $x = 1 + \sqrt{2}$. Similarly, the relations

$$n \leq \frac{n - \sqrt{20 - 3n^2}}{2} < n + 1$$

are easily seen to be impossible (for instance, the inequality on the left would yield

$$n \leq -\sqrt{20 - 3n^2} \leq -\sqrt{8} < -2,$$

a contradiction). We conclude that the equation has a unique solution, namely $x = 1 + \sqrt{2}$.

Example 10.7. Find a nonzero polynomial $P(x, y)$ such that $P(\lfloor x \rfloor, \lfloor 2x \rfloor) = 0$ for all real numbers x.

Solution. The key point is that $\lfloor 2x \rfloor$ is quite close to $2\lfloor x \rfloor$. Indeed, for all x we have

$$\lfloor 2x \rfloor = 2\lfloor x \rfloor + \lfloor 2\{x\} \rfloor,$$

so that $\lfloor 2x \rfloor - 2\lfloor x \rfloor \in \{0, 1\}$. It suffices therefore to find a polynomial $P(x, y)$ such that $P(x, y) = 0$ whenever $y = 2x$ or $y = 2x + 1$. Clearly $P(x, y) = (y - 2x)(y - 2x - 1)$ is such a polynomial.

Example 10.8. Find the smallest value that $\lfloor x \rfloor$ can take, knowing that

$$\lfloor x^2 \rfloor - \lfloor x \rfloor^2 = 100.$$

Solution. Let $n = \lfloor x \rfloor$. Since

$$x^2 = (n + \{x\})^2 = n^2 + 2n\{x\} + \{x\}^2,$$

the equation is equivalent to

$$\lfloor 2n\{x\} + \{x\}^2 \rfloor = 100,$$

or equivalently

$$100 \leq 2n\{x\} + \{x\}^2 < 101.$$

The inequality on the left combined with $\{x\} < 1$ yields $99 < 2n$ and so $n \geq 50$ (since n is an integer). Let us see whether $n = 50$ works. We need to check whether we can find $y \in [0, 1)$ such that

$$100 \leq 100y + y^2 < 101.$$

This suggests taking y close to 1. Set $y = 1 - z$. The conditions become $100z < (1-z)^2$ and $1 + 100z > (1-z)^2$. We see that taking $z = \frac{1}{200}$ (or something smaller) both inequalities are satisfied. Thus the answer of the problem is 50.

Example 10.9. Let $a_n = \lfloor \sqrt{n} + \frac{1}{2} \rfloor$. Compute $\sum_{n=1}^{2004} \frac{1}{a_n}$.

Solution. We note that $a_n = k$ if and only if

$$k \leq \sqrt{n} + \frac{1}{2} < k+1, \quad \text{or} \quad k^2 - k + \frac{1}{4} \leq n < k^2 + k + \frac{1}{4}.$$

Since n and k are integers, the latter happens if and only if $k^2 - k + 1 \leq n \leq k^2 + k$. Let $x_k = k^2 - k$, then $x_{k+1} = k^2 + k$ and so we obtain that $a_n = k$ if and only if $x_k < n \leq x_{k+1}$, and this happens for precisely $x_{k+1} - x_k = 2k$ values of n. This being said and taking into account that $x_{45} < 2004 < x_{46}$, we obtain

$$\sum_{n=1}^{2004} \frac{1}{a_n} = \sum_{k=1}^{44} \sum_{x_k < n \leq x_{k+1}} \frac{1}{k} + \sum_{x_{45} < n \leq 2004} \frac{1}{45} =$$

$$44 \cdot 2 + \frac{24}{45} = 88 + \frac{8}{15}.$$

Example 10.10. Prove that for all real numbers x we have

$$\left\lfloor \frac{x}{3} \right\rfloor + \left\lfloor \frac{x+2}{6} \right\rfloor + \left\lfloor \frac{x+4}{6} \right\rfloor = \left\lfloor \frac{x}{2} \right\rfloor + \left\lfloor \frac{x+3}{6} \right\rfloor.$$

Solution. Write the equality as

$$\left\lfloor\frac{x}{3}\right\rfloor+\left\lfloor\frac{x}{6}+\frac{1}{3}\right\rfloor+\left\lfloor\frac{x}{6}+\frac{2}{3}\right\rfloor=\left\lfloor\frac{x}{2}\right\rfloor+\left\lfloor\frac{x}{6}+\frac{1}{2}\right\rfloor.$$

By Hermite's identity we have

$$\left\lfloor\frac{x}{6}\right\rfloor+\left\lfloor\frac{x}{6}+\frac{1}{3}\right\rfloor+\left\lfloor\frac{x}{6}+\frac{2}{3}\right\rfloor=\left\lfloor 3\cdot\frac{x}{6}\right\rfloor=\left\lfloor\frac{x}{2}\right\rfloor.$$

It suffices therefore to prove that

$$\left\lfloor\frac{x}{3}\right\rfloor+\left\lfloor\frac{x}{2}\right\rfloor-\left\lfloor\frac{x}{6}\right\rfloor=\left\lfloor\frac{x}{2}\right\rfloor+\left\lfloor\frac{x}{6}+\frac{1}{2}\right\rfloor.$$

But this follows again from Hermite's identity, since

$$\left\lfloor\frac{x}{6}\right\rfloor+\left\lfloor\frac{x}{6}+\frac{1}{2}\right\rfloor=\left\lfloor 2\cdot\frac{x}{6}\right\rfloor=\left\lfloor\frac{x}{3}\right\rfloor.$$

Example 10.11. Prove that for all positive integers n,

$$\left\lfloor\frac{n+2^0}{2^1}\right\rfloor+\left\lfloor\frac{n+2^1}{2^2}\right\rfloor+\cdots+\left\lfloor\frac{n+2^{n-1}}{2^n}\right\rfloor=n.$$

Solution. Write the equality as

$$\left\lfloor\frac{n}{2}+\frac{1}{2}\right\rfloor+\left\lfloor\frac{n}{4}+\frac{1}{2}\right\rfloor+\ldots+\left\lfloor\frac{n}{2^n}+\frac{1}{2}\right\rfloor=n.$$

Using Hermite's identity in the form

$$\left\lfloor x+\frac{1}{2}\right\rfloor=\lfloor 2x\rfloor-\lfloor x\rfloor,$$

we obtain

$$\left\lfloor\frac{n}{2}+\frac{1}{2}\right\rfloor+\left\lfloor\frac{n}{4}+\frac{1}{2}\right\rfloor+\ldots+\left\lfloor\frac{n}{2^n}+\frac{1}{2}\right\rfloor=$$

$$n-\left\lfloor\frac{n}{2}\right\rfloor+\left\lfloor\frac{n}{2}\right\rfloor-\left\lfloor\frac{n}{4}\right\rfloor+\ldots+\left\lfloor\frac{n}{2^{n-1}}\right\rfloor-\left\lfloor\frac{n}{2^n}\right\rfloor=n-\left\lfloor\frac{n}{2^n}\right\rfloor.$$

Now, it is easy to see that $2^n > n$ for all n (for instance by expanding

$$2^n=(1+1)^n=1+n+\ldots$$

using the binomial formula). Thus $\lfloor\frac{n}{2^n}\rfloor=0$ and we are done.

Example 10.12. Prove that for all positive integers n we have

$$\lfloor\sqrt{n}+\sqrt{n+1}\rfloor=\lfloor\sqrt{4n+1}\rfloor.$$

Solution. Let us denote $\lfloor\sqrt{4n+1}\rfloor = k$, so that $k^2 \leq 4n+1 < (k+1)^2$. We need to prove that
$$k \leq \sqrt{n} + \sqrt{n+1} < k+1,$$
or equivalently, by squaring the inequality
$$k^2 \leq 2n+1 + 2\sqrt{n^2+n} < (k+1)^2.$$

The inequality on the left is immediate, since
$$2n+1 + 2\sqrt{n^2+n} > 2n+1+2n = 4n+1 \geq k^2.$$

The inequality on the right is proved in a similar, though a bit more subtle way
$$2n+1 + 2\sqrt{n^2+n} < 2n+1 + 2\left(n+\frac{1}{2}\right) = 4n+2 \leq (k+1)^2.$$

Note that the last inequality follows from $4n+1 < (k+1)^2$, since $4n+1$ and $(k+1)^2$ are integers.

Example 10.13. Solve the equation
$$2x^2 + \frac{\lfloor x \rfloor^2}{2} + 2\{x\}^2 = 12x - 15.$$

Solution. Let
$$\lfloor x \rfloor = n \quad \text{and} \quad \{x\} = y.$$
The equation is equivalent to
$$2(n+y)^2 + \frac{n^2}{2} + 2y^2 = 12(n+y) - 15$$
or equivalently, by expanding, multiplying by 2 and rearranging terms
$$5n^2 + 8n(y-3) + 8y^2 - 24y + 30 = 0.$$
Treating this as a quadratic equation in n, its discriminant is
$$\Delta = 64(y-3)^2 - 20(8y^2 - 24y + 30) =$$
$$-96y^2 + 96y - 24 = -24(4y^2 - 4y + 1) = -24(2y-1)^2 \leq 0.$$
Since the equation has real roots, we deduce that $2y = 1$, so $y = \frac{1}{2}$, and
$$n = -\frac{8(y-3)}{10} = 2.$$
Thus
$$x = n + y = 2 + \frac{1}{2} = \frac{5}{2}$$
is the unique solution of the equation.

Example 10.14. Let n be a positive integer. Find the integer part of $\sum_{k=1}^{2n} \sqrt{n^2 + k}$.

Solution. Each term $\sqrt{n^2 + k}$ is quite close to n, so we will rather focus on the difference

$$\sum_{k=1}^{2n} \left(\sqrt{n^2 + k} - n \right) = \sum_{k=1}^{2n} \frac{k}{n + \sqrt{n^2 + k}}.$$

Next, observe that for $1 \leq k \leq 2n$ we have

$$\frac{k}{2n+1} < \frac{k}{n + \sqrt{n^2 + k}} < \frac{k}{2n},$$

since

$$n < \sqrt{n^2 + k} < n + 1.$$

Adding up these relations, we end up with

$$\frac{2n(2n+1)}{2(2n+1)} < \sum_{k=1}^{2n} \left(\sqrt{n^2 + k} - n \right) < \frac{2n(2n+1)}{2 \cdot 2n},$$

that is

$$n + \sum_{k=1}^{2n} n < \sum_{k=1}^{2n} \sqrt{n^2 + k} < n + \frac{1}{2} + \sum_{k=1}^{2n} n.$$

This clearly implies the inequality

$$n + 2n^2 < \sum_{k=1}^{2n} \sqrt{n^2 + k} < n + 2n^2 + 1$$

and shows that the answer of the problem is $2n^2 + n$.

Example 10.15. Let x be a real number. Prove that x is an integer if and only if for all positive integers n we have

$$\lfloor x \rfloor + \lfloor 2x \rfloor + \lfloor 3x \rfloor + \ldots + \lfloor nx \rfloor = \frac{n(\lfloor x \rfloor + \lfloor nx \rfloor)}{2}.$$

Solution. If x is an integer, then the desired equality comes down to

$$1 + 2 + \ldots + n = \frac{n(n+1)}{2},$$

which is well-known (and follows by adding up the relations $k + (n - k) = n$ for $0 \leq k \leq n$). The converse is more delicate. Suppose that x satisfies the relation in the statement of the problem for all n. Comparing the relations for n and $n-1$ and subtracting them yields

$$\lfloor nx \rfloor = \frac{\lfloor x \rfloor + n\lfloor nx \rfloor - (n-1)\lfloor (n-1)x \rfloor}{2},$$

which can be written as

$$(n-2)\lfloor nx \rfloor + \lfloor x \rfloor = (n-1)\lfloor (n-1)x \rfloor.$$

Since $\lfloor nx \rfloor = n\lfloor x \rfloor + \lfloor n\{x\} \rfloor$ and similarly with $n-1$ instead of n, we obtain the equivalent relation

$$(n-2)\lfloor n\{x\} \rfloor = (n-1)\lfloor (n-1)\{x\} \rfloor.$$

Suppose that x is not an integer, so $\{x\} > 0$. Then for $n > \frac{1}{\{x\}} + 1$ we have $\lfloor (n-1)\{x\} \rfloor > 0$. On the other hand, the previous relation shows that $n-2$ divides $\lfloor (n-1)\{x\} \rfloor > 0$, hence we must have $\lfloor (n-1)\{x\} \rfloor \geq n-2$. Thus $(n-1)\{x\} \geq n-2$, and this for all $n > 1 + \frac{1}{\{x\}}$. But then

$$\{x\} \geq 1 - \frac{1}{n-1} \quad \text{so} \quad n \leq 1 + \frac{1}{1-\{x\}}$$

for all $n > 1 + \frac{1}{\{x\}}$. Since this is clearly impossible (it would imply that there are only finitely many positive integers), the result follows.

Example 10.16. A sequence $(a_n)_{n \geq 1}$ of positive integers is defined by

$$a_n = \left\lfloor n + \sqrt{n} + \frac{1}{2} \right\rfloor.$$

Determine the positive integers that occur in the sequence.

Solution. Writing down the first few terms we obtain $2, 3, 5, 6, 7, 8, 10, \ldots$ and we see that the sequence seems to skip precisely the perfect squares. Let us prove this. First, it is clear that the sequence is increasing. Suppose that $a_n = k$. Then the sequence skips the value $k+1$ if and only if $a_{n+1} \geq k+2$, which happens if and only if

$$n + 1 + \sqrt{n+1} + \frac{1}{2} \geq k + 2.$$

Indeed, for a real number x and an **integer** r we have $\lfloor x \rfloor \geq r$ if and only if $x \geq r$ (this follows straight from the definition of the floor function). Let $m = k - n$. Then the previous inequality is equivalent to

$$n + 1 \geq \left(m + \frac{1}{2}\right)^2 = m^2 + m + \frac{1}{4}.$$

Since m and n are integers, this is in turn equivalent to $n \geq m^2 + m$. On the other hand, the condition $a_n = k$ can be written as

$$k \leq n + \sqrt{n} + \frac{1}{2} < k + 1$$

and in terms of m this becomes

$$\left(m - \frac{1}{2}\right)^2 < n < \left(m + \frac{1}{2}\right)^2.$$

Expanding again and taking care of the fact that m and n are integers, we obtain the equivalent condition

$$m^2 - m < n \leq m^2 + m.$$

Combining this with the previously established inequality $n \geq m^2 + m$, we see that we must have $n = m^2 + m$, and in this case all relations are satisfied. But then $k = n + m = m^2 + 2m$ and $k + 1 = (m+1)^2$. Hence we proved that the sequence skips a positive integer $k+1$ if and only if there is a nonnegative integer m such that $k+1 = (m+1)^2$. This clearly shows that the set of values taken by the sequence $(a_n)_{n \geq 1}$ is exactly the set of all non perfect squares.

Example 10.17. Prove that for all primes p we have

$$\sum_{k=1}^{p-1} \left\lfloor \frac{k^3}{p} \right\rfloor = \frac{(p^2-1)(p-2)}{4}.$$

Solution. This is rather tricky. Let us fix $1 \leq k < p$ and focus on

$$\left\lfloor \frac{k^3}{p} \right\rfloor + \left\lfloor \frac{(p-k)^3}{p} \right\rfloor.$$

Since

$$k^3 + (p-k)^3 = p(k^2 - k(p-k) + (p-k)^2)$$

is a multiple of p, we can write $(p-k)^3 = np - k^3$ for some integer n. Thus

$$\left\lfloor \frac{(p-k)^3}{p} \right\rfloor = \left\lfloor n - \frac{k^3}{p} \right\rfloor = n + \left\lfloor -\frac{k^3}{p} \right\rfloor.$$

On the other hand, if $\frac{k^3}{p} \in [N, N+1)$ for some integer N, then $-\frac{k^3}{p} \in (-N-1, -N]$. Moreover, we cannot have $-\frac{k^3}{p} = -N$, since then p would divide k^3 and this contradicts the hypothesis that p is a prime and $1 \leq k < p$ (thus k is relatively prime to p). Thus we must have

$$\left\lfloor -\frac{k^3}{p} \right\rfloor = -N - 1 = -1 - \left\lfloor \frac{k^3}{p} \right\rfloor.$$

The floor function

We conclude that

$$\left\lfloor \frac{(p-k)^3}{p} \right\rfloor + \left\lfloor \frac{k^3}{p} \right\rfloor = -1 + n = -1 + \frac{k^3 + (p-k)^3}{p}.$$

Adding up these relations for $k = 1, 2, ..., p-1$ and using the identity

$$1^3 + 2^3 + ... + n^3 = \frac{n^2(n+1)^2}{4}$$

yields

$$2\sum_{k=1}^{p-1} \left\lfloor \frac{k^3}{p} \right\rfloor = -(p-1) + 2\sum_{k=1}^{p-1} \frac{k^3}{p} =$$

$$-(p-1) + 2\frac{(p-1)^2 p^2}{4p} = \frac{p(p-1)^2}{2} - (p-1) = (p-1) \cdot \frac{p^2 - p - 2}{2}.$$

Finally, dividing by 2 and noticing that $p^2 - p - 2 = (p+1)(p-2)$ yields the desired result.

Example 10.18. Prove that for all real numbers x we have the following inequality

$$\sum_{k=1}^{n} \frac{\lfloor kx \rfloor}{k} \leq \lfloor nx \rfloor.$$

Solution. This problem is very challenging. We will prove the result by strong induction, the case $n = 1$ being clear. Suppose that the statement holds up to n and let us prove it for n. Let

$$S_j = \sum_{k=1}^{j} \frac{\lfloor kx \rfloor}{k} = S_{j-1} + \frac{\lfloor jx \rfloor}{j}$$

so that

$$j(S_j - S_{j-1}) = \lfloor jx \rfloor$$

for $j \geq 2$. Adding up these relations yields

$$nS_n = S_1 + ... + S_{n-1} + \sum_{k=1}^{n} \lfloor kx \rfloor,$$

hence using the inductive hypothesis

$$nS_n \leq \sum_{k=1}^{n-1} \lfloor kx \rfloor + \sum_{k=1}^{n} \lfloor kx \rfloor.$$

The tricky step is to prove that the last quantity is at most $n\lfloor nx \rfloor$. In order to do this, we observe that

$$\sum_{k=1}^{n-1} \lfloor kx \rfloor + \sum_{k=1}^{n} \lfloor kx \rfloor = \sum_{k=1}^{n} \lfloor (n-k)x \rfloor + \sum_{k=1}^{n} \lfloor kx \rfloor =$$

$$\sum_{k=1}^{n} \left(\lfloor (n-k)x \rfloor + \lfloor kx \rfloor \right),$$

so it is enough to prove that

$$\lfloor (n-k)x \rfloor + \lfloor kx \rfloor \leq \lfloor nx \rfloor$$

for all k, n and x. But this is simply the inequality $\lfloor a \rfloor + \lfloor b \rfloor \leq \lfloor a + b \rfloor$ for $a = (n-k)x$ and $b = kx$. This finishes the proof.

11 Taking advantage of symmetry

When proving inequalities or solving systems of equations, one faces quite often symmetric polynomial expressions. These are polynomial expressions $f(a, b, c, ...)$ in several variables, which do no change when permuting the variables $a, b, c, ...$. For example, $a^2+b^2+c^2$, $ab+bc+ca+(a+b+c)^2$ are symmetric polynomial expressions in a, b, c, since they obviously do not change when permuting a, b, c.

If $x_1, x_2, ..., x_n$ are variables, then there is a class of fundamental symmetric expressions in $x_1, x_2, ..., x_n$. They are given by the fundamental symmetric sums (with $k = 1, 2, ..., n$)

$$\sigma_k(x_1, ..., x_n) = \sum_{1 \leq i_1 < i_2 < ... < i_k \leq n} x_{i_1} x_{i_2} ... x_{i_k},$$

the summation being taken over all k-tuples $(i_1, ..., i_k)$ with $1 \leq i_1 < ... < i_k \leq n$. For instance

$$\sigma_1(x_1, ..., x_n) = x_1 + x_2 + ... + x_n,$$

$$\sigma_2(x_1, ..., x_n) = \sum_{i<j} x_i x_j = x_1 x_2 + ... + x_1 x_n + x_2 x_3 + ... + x_2 x_n + ... + x_{n-1} x_n,$$

$$\sigma_n = x_1 x_2 ... x_n.$$

The fundamental relation between these sums is (we write simply σ_k instead of $\sigma_k(x_1, ..., x_n)$)

$$(t + x_1)(t + x_2)...(t + x_n) = t^n + \sigma_1 t^{n-1} + \sigma_2 t^{n-2} + ... + \sigma_n,$$

which follows by brutally expanding the left hand-side. This identity holds for all t and all $x_1, x_2, ..., x_n$. Taking $t = -x_i$, the left hand-side becomes 0, hence we have just proved part a) of the following fundamental

Theorem 11.1. *(Vieta's relations)* a) $x_1, x_2, ..., x_n$ *are solutions of the equation*

$$t^n - \sigma_1 t^{n-1} + \sigma_2 t^{n-2} - ... + (-1)^n \sigma_n = 0,$$

where $\sigma_k = \sigma_k(x_1, ..., x_n)$.

b) *If* $x_1, x_2, ..., x_n$ *are the solutions of the equation*

$$a_n t^n + a_{n-1} t^{n-1} + ... + a_0 = 0,$$

with $a_n \neq 0$, *then we have Vieta's formulae*

$$\sigma_k(x_1, x_2, ..., x_n) = (-1)^k \frac{a_{n-k}}{a_n}$$

for all $1 \leq k \leq n$.

Part b) is proved similarly, by observing that by hypothesis we have

$$t^n + \frac{a_n}{a_{n-1}}t^{n-1} + ... + \frac{a_0}{a_n} = (t-x_1)(t-x_2)...(t-x_n),$$

expanding the right hand-side to get $t^n - \sigma_1 t^{n-1} + ... + (-1)^n \sigma_n$ and identifying the coefficients of t^k for all $1 \le k \le n$.

The following theorem is more difficult and one of the most important theorems in algebra:

Theorem 11.2. *Let $p(x_1, x_2, ..., x_n)$ be a symmetric polynomial expression in $x_1, x_2, ..., x_n$. Then we can find a polynomial expression $q(y_1, y_2, ..., y_n)$ such that*

$$p(x_1, x_2, ..., x_n) = q(\sigma_1, \sigma_2, ..., \sigma_n),$$

where $\sigma_k = \sigma_k(x_1, x_2, ..., x_n)$.

Stated less formally, any symmetric polynomial expression in $x_1, x_2, ..., x_n$ is a polynomial expression in the fundamental symmetric sums $\sigma_1, \sigma_2, ..., \sigma_n$ associated to $x_1, x_2, ..., x_n$. The proof of the theorem is beyond the scope of this introductory book, but we will see many explicit examples of how this theorem works. A very important class of symmetric polynomials is given by the polynomials $x_1^k + x_2^k + ... + x_n^k$, with $k \ge 1$. The following fundamental theorem makes the previous theorem more explicit for these polynomials. It is crucial that one have all the roots in the following statement.

Theorem 11.3. *(Newton's relations) Let $x_1, x_2, ..., x_n$ be the solutions of the equation $x^n + a_1 x^{n-1} + ... + a_n = 0$, where $a_1, a_2, ..., a_n$ are given real numbers. Define*

$$S_k = x_1^k + x_2^k + ... + x_n^k.$$

Then for all $k \in \{1, 2, ..., n\}$ we have

$$S_k + a_1 S_{k-1} + ... + a_{k-1} S_1 + k a_k = 0,$$

while for all $k > n$ we have

$$S_k + a_1 S_{k-1} + ... + a_n S_{k-n} = 0.$$

We will give only the proof of the second part (which is the easier one) of the theorem, since we will constantly use this kind of argument in the examples. Since x_i are solutions of the equation $x^n + a_1 x^{n-1} + ... + a_n = 0$, we can write

$$x_i^n + a_1 x_i^{n-1} + ... + a_n = 0, \quad \text{hence} \quad x_i^{n+j} + a_1 x_i^{n+j-1} + ... + a_n x_i^j = 0$$

for all $j \geq 0$. By adding these relations for $i = 1, 2, ..., n$, we obtain

$$S_{n+j} + a_1 S_{n+j-1} + ... + a_n S_j = 0$$

for $j \geq 0$. Thus

$$S_k + a_1 S_{k-1} + ... + a_n S_{k-n} = 0$$

for $k \geq n$, which is the second part of the theorem (and a tiny bit of the first part).

Let us move on to examples now!

Example 11.1. Express the following polynomials in terms of $a + b + c$, $ab + bc + ca$ and abc:
 a) $a^2 + b^2 + c^2$.
 b) $a^3 + b^3 + c^3$.
 c) $ab(a + b) + bc(b + c) + ca(c + a)$.
 d) $a^2(b + c) + b^2(c + a) + c^2(a + b)$.

Solution. a) This follows directly from the identity

$$(a + b + c)^2 = a^2 + b^2 + c^2 + 2(ab + bc + ca),$$

which gives

$$a^2 + b^2 + c^2 = (a + b + c)^2 - 2(ab + bc + ca).$$

b) We use the identity

$$a^3 + b^3 + c^3 - 3abc = (a + b + c)(a^2 + b^2 + c^2 - ab - bc - ca).$$

Combining this with the result established in a), we obtain

$$a^3 + b^3 + c^3 = 3abc + (a + b + c)((a + b + c)^2 - 3(ab + bc + ca)).$$

Here is another method to solve this: a, b, c are solutions of the equation

$$x^3 - (a + b + c)x^2 + (ab + bc + ca)x - abc = 0$$

so adding up these relations for $x = a, b, c$ yields

$$a^3 + b^3 + c^3 - (a + b + c)(a^2 + b^2 + c^2) + (ab + bc + ca)(a + b + c) - 3abc = 0,$$

which combined with a) yields the desired result.

c) We have

$$ab(a+b)+bc(b+c)+ca(c+a) = ab(a+b+c-c)+bc(a+b+c-a)+ca(a+b+c-b) =$$

$$(ab + bc + ca)(a + b + c) - 3abc.$$

d) The easiest way is to note that

$$a^2(b+c) + b^2(c+a) + c^2(a+b) = a^2b + ab^2 + b^2c + bc^2 + c^2a + a^2c$$
$$= ab(a+b) + bc(b+c) + ca(c+a)$$

and then to use c). One could also argue as follows: write

$$a^2(b+c) + b^2(c+a) + c^2(a+b) = a^2(a+b+c-a) + b^2(a+b+c-b) + c^2(a+b+c-c) =$$
$$(a^2 + b^2 + c^2)(a+b+c) - a^3 - b^3 - c^3,$$

and then use a) and b). However, simplifying the resulting expression is much more tedious than in the previous approach, which gives the final result directly.

Example 11.2. Write $x_1^2 + x_2^2 + ... + x_n^2$ as a polynomial expression in the symmetric fundamental sums $\sigma_1, ..., \sigma_n$ associated to $x_1, x_2, ..., x_n$. Do the same with $x_1^3 + x_2^3 + ... + x_n^3$ if $n \geq 3$.

Solution. Start by expanding

$$\sigma_1^2 = (x_1 + x_2 + ... + x_n)^2 = x_1^2 + x_2^2 + ... + x_n^2 + 2\sum_{i<j} x_i x_j = x_1^2 + x_2^2 + ... + x_n^2 + 2\sigma_2,$$

hence

$$x_1^2 + x_2^2 + ... + x_n^2 = \sigma_1^2 - 2\sigma_2.$$

Note that we could have also used Newton's relations, which would have given this directly. For the second polynomial, it is more convenient to use Newton's relations: if $x_1, ..., x_n$ are solutions of the equation $x^n + a_1 x^{n-1} + ... + a_n = 0$, then theorem 11.3 gives

$$x_1^3 + x_2^3 + ... + x_n^3 + a_1(x_1^2 + ... + x_n^2) + a_2(x_1 + ... + x_n) + 3a_3 = 0.$$

On the other hand, Vieta's relations yield (with $\sigma_k = \sigma_k(x_1, ..., x_n)$)

$$a_1 = -\sigma_1, \quad a_2 = \sigma_2, \quad a_3 = -\sigma_3.$$

Thus

$$x_1^3 + x_2^3 + ... + x_n^3 = \sigma_1(x_1^2 + ... + x_n^2) - \sigma_2(x_1 + ... + x_n) + 3\sigma_3$$

and using that $x_1^2 + ... + x_n^2 = \sigma_1^2 - 2\sigma_2$, we finally obtain

$$x_1^3 + x_2^3 + ... + x_n^3 = \sigma_1^3 - 3\sigma_1 \sigma_2 + 3\sigma_3.$$

Example 11.3. Pairwise distinct real numbers satisfy
$$a(a^2 - abc) = b(b^2 - abc) = c(c^2 - abc).$$
Prove that $a + b + c = 0$.

Solution. Let $p = abc$ and let d be the common value of the expressions in the statement of the problem. Then by hypothesis a, b, c are the different solutions of the equation $t^3 - pt - d = 0$. Using Vieta's formulae, we obtain $a + b + c = 0$, since the coefficient of t^2 in $t^3 - pt - d$ is 0.

Example 11.4. Let x_1 and x_2 be the solutions of the equation $x^2 + 3x + 1 = 0$. Compute
$$\left(\frac{x_1}{x_2 + 1}\right)^2 + \left(\frac{x_2}{x_1 + 1}\right)^2.$$

Solution. We start by replacing x_1^2 by $-3x_1 - 1$ and x_2^2 by $-3x_2 - 1$. In particular
$$(x_2 + 1)^2 = x_2^2 + 2x_2 + 1 = -3x_2 - 1 + 2x_2 + 1 = -x_2$$
and similarly with x_2 replaced by x_1. We obtain therefore
$$\left(\frac{x_1}{x_2 + 1}\right)^2 + \left(\frac{x_2}{x_1 + 1}\right)^2 = \frac{3x_1 + 1}{x_2} + \frac{3x_2 + 1}{x_1} =$$
$$= \frac{3x_1^2 + x_1 + 3x_2^2 + x_2}{x_1 x_2}.$$
We use next Vieta's relations to get $x_1 + x_2 = -3$ and $x_1 x_2 = 1$. Thus
$$x_1^2 + x_2^2 = (x_1 + x_2)^2 - 2x_1 x_2 = 7$$
and finally
$$\left(\frac{x_1}{x_2 + 1}\right)^2 + \left(\frac{x_2}{x_1 + 1}\right)^2 = \frac{21 - 3}{1} = 18.$$

Example 11.5. Let a, b, c be real numbers such that $a + b + c$, $ab + bc + ca$ and abc are all positive. Prove that a, b, c are positive.

Solution. The numbers a, b, c are roots of $x^3 - \sigma_1 x^2 + \sigma_2 x - \sigma_3$, where
$$\sigma_1 = a + b + c, \quad \sigma_2 = ab + bc + ca, \quad \sigma_3 = abc.$$
It suffices therefore to prove that if $\sigma_1, \sigma_2, \sigma_3 > 0$, then any real root of the previous polynomial is positive. Suppose that $x = -y$ is a root, with $y \geq 0$. Then
$$-y^3 - \sigma_1 y^2 - \sigma_2 y - \sigma_3 = 0.$$

But this is clearly impossible, since each term in the left hand-side is nonpositive, while the last term is negative by assumption. Hence the left hand-side is negative. This contradiction shows that any solution is positive, and so $a, b, c > 0$.

Example 11.6. Let a, b, c be the solutions of the equation $x^3 - x - 1 = 0$. Compute
$$\frac{1-a}{1+a} + \frac{1-b}{1+b} + \frac{1-c}{1+c}.$$

Solution. We start by simplifying a little bit the given expression:
$$\frac{1-a}{1+a} + \frac{1-b}{1+b} + \frac{1-c}{1+c} = \frac{-(1+a)+2}{1+a} + \frac{-(1+b)+2}{1+b} + \frac{-(1+c)+2}{1+c} =$$
$$2\left(\frac{1}{1+a} + \frac{1}{1+b} + \frac{1}{1+c}\right) - 3.$$

Next, we compute
$$\frac{1}{1+a} + \frac{1}{1+b} + \frac{1}{1+c} = \frac{(1+a)(1+b) + (1+b)(1+c) + (1+c)(1+a)}{(1+a)(1+b)(1+c)}$$

and we deal separately with the numerator and the denominator. We have
$$(1+a)(1+b) + (1+b)(1+c) + (1+c)(1+a) =$$
$$1 + a + b + ab + 1 + b + c + bc + 1 + c + a + ca =$$
$$3 + 2(a+b+c) + ab + bc + ca = 3 - 1 = 2,$$

where we used Vieta's relations
$$a + b + c = 0, \quad ab + bc + ca = -1.$$

Next, since
$$x^3 - x - 1 = (x-a)(x-b)(x-c)$$

for all x, choosing $x = -1$ yields
$$-1 = -(1+a)(1+b)(1+c), \quad \text{hence} \quad (1+a)(1+b)(1+c) = 1.$$

Combining all this yields
$$\frac{1-a}{1+a} + \frac{1-b}{1+b} + \frac{1-c}{1+c} = 2 \cdot 2 - 3 = 1.$$

Here is an alternative way of computing $\frac{1}{1+a} + \frac{1}{1+b} + \frac{1}{1+c}$. We will find a third degree equation having as solutions $\frac{1}{1+a}, \frac{1}{1+b}$ and $\frac{1}{1+c}$. Suppose that x

is a solution of the equation $x^3 - x - 1 = 0$ and let $y = \frac{1}{1+x}$. Then $x = \frac{1}{y} - 1$ and so
$$\left(\frac{1}{y} - 1\right)^3 - \left(\frac{1}{y} - 1\right) - 1 = 0.$$
Expanding everything and multiplying by $-y^3$ yields the equivalent equation
$$y^3 - 2y^2 + 3y - 1 = 0.$$
This is the equation whose solutions are $\frac{1}{1+a}$, $\frac{1}{1+b}$ and $\frac{1}{1+c}$. Using Vieta's formulae we obtain
$$\frac{1}{1+a} + \frac{1}{1+b} + \frac{1}{1+c} = 2.$$

Example 11.7. A polynomial $p(x) = x^n + a_1 x^{n-1} + ... + a_{n-1} x + 1$ has nonnegative coefficients and all of its roots are real numbers. Prove that $p(2) \geq 3^n$.

Solution. Let $x_1, x_2, ..., x_n$ be the roots of p. Since $x_i^n + a_1 x_i^{n-1} + ... + 1 = 0$ and since $a_1, a_2, ..., a_{n-1}$ are all nonnegative, it follows that $x_i < 0$ for all i. Write $x_i = -y_i$ for some $y_i > 0$. Vieta's relations yield
$$y_1 y_2 ... y_n = (-1)^n x_1 x_2 ... x_n = 1$$
and we need to prove that
$$(2 + y_1)(2 + y_2)...(2 + y_n) \geq 3^n.$$
This is an easy application of the AM-GM inequality: it suffices to multiply the inequalities $2 + y_i \geq 3\sqrt[3]{y_i}$, which follow straight from the AM-GM inequality.

Example 11.8. Let x_1, x_2, x_3 be the roots of the polynomial $x^3 + 3x + 1$. Compute
$$\frac{x_1^2}{(5x_2 + 1)(5x_3 + 1)} + \frac{x_2^2}{(5x_1 + 1)(5x_3 + 1)} + \frac{x_3^2}{(5x_1 + 1)(5x_2 + 1)}.$$

Solution. We compute rather directly
$$\frac{x_1^2}{(5x_2 + 1)(5x_3 + 1)} + \frac{x_2^2}{(5x_1 + 1)(5x_3 + 1)} + \frac{x_3^2}{(5x_1 + 1)(5x_2 + 1)} =$$
$$= \frac{x_1^2(5x_1 + 1) + x_2^2(5x_2 + 1) + x_3^2(5x_3 + 1)}{(5x_1 + 1)(5x_2 + 1)(5x_3 + 1)}.$$

Next, we compute separately the numerator and the denominator of the previous fraction. Observe that
$$x_1^2(5x_1 + 1) = 5x_1^3 + x_1^2 = 5(-3x_1 - 1) + x_1^2 = x_1^2 - 15x_1 - 5$$

and similarly with x_2 and x_3 instead of x_1. Hence
$$x_1^2(5x_1+1) + x_2^2(5x_2+1) + x_3^2(5x_3+1) =$$
$$(x_1^2 + x_2^2 + x_3^2) - 15(x_1 + x_2 + x_3) - 15.$$
From Vieta's relations we obtain $x_1 + x_2 + x_3 = 0$ and $x_1 x_2 + x_2 x_3 + x_3 x_1 = 3$. Thus
$$x_1^2 + x_2^2 + x_3^2 = (x_1+x_2+x_3)^2 - 2(x_1 x_2 + x_2 x_3 + x_3 x_1) = -6$$
and
$$x_1^2(5x_1+1) + x_2^2(5x_2+1) + x_3^2(5x_3+1) = -21.$$
Next, we deal with the denominator
$$(5x_1+1)(5x_2+1)(5x_3+1) = 5^3 \left(x_1 + \frac{1}{5}\right)\left(x_2 + \frac{1}{5}\right)\left(x_3 + \frac{1}{5}\right).$$
Since
$$x^3 + 3x + 1 = (x-x_1)(x-x_2)(x-x_3)$$
for all x, by choosing $x = -\frac{1}{5}$ we obtain
$$-\left(x_1 + \frac{1}{5}\right)\left(x_2 + \frac{1}{5}\right)\left(x_3 + \frac{1}{5}\right) = -\frac{1}{125} - \frac{3}{5} + 1.$$
Combining these relations, we obtain that
$$(5x_1+1)(5x_2+1)(5x_3+1) = -49$$
and finally
$$\frac{x_1^2}{(5x_2+1)(5x_3+1)} + \frac{x_2^2}{(5x_1+1)(5x_3+1)} + \frac{x_3^2}{(5x_1+1)(5x_2+1)} = \frac{21}{49} = \frac{3}{7}.$$

Example 11.9. Let a, b, c, d, e, f be real numbers such that all roots of the polynomial
$$p(x) = x^8 - 4x^7 + 7x^6 + ax^5 + bx^4 + cx^3 + dx^2 + ex + f$$
are real numbers. Find all possible values of f.

Solution. Let $x_1, x_2, ..., x_8$ be the roots of the polynomial p and observe that Vieta's relations yield
$$x_1 + x_2 + ... + x_8 = 4, \quad \sum_{1 \le i < j \le 8} x_i x_j = 7.$$

Since
$$(x_1 + x_2 + \ldots + x_8)^2 = x_1^2 + x_2^2 + \ldots + x_8^2 + 2\sum_{i<j} x_i x_j,$$

it follows that $x_1^2 + x_2^2 + \ldots + x_8^2 = 2$. On the other hand, the Cauchy-Schwarz inequality gives
$$x_1^2 + x_2^2 + \ldots + x_8^2 \geq \frac{(x_1 + x_2 + \ldots + x_8)^2}{8}.$$

Taking into account the previous relations, this becomes $2 \geq 2$ and so it must be an equality. But that forces $x_1 = x_2 = \ldots = x_8$ and since $x_1 + x_2 + \ldots + x_8 = 4$, we must have $x_1 = x_2 = \ldots = x_8 = \frac{1}{2}$. Finally, Vieta's relations yield
$$f = x_1 x_2 \ldots x_8 = \frac{1}{2^8} = \frac{1}{256}.$$

Example 11.10. If a, b, c are real numbers such that
$$a + b + c = 1 \quad \text{and} \quad \frac{1}{a} + \frac{1}{b} + \frac{1}{c} = 2,$$

find
$$\left(\frac{1}{a} - 1\right)\left(\frac{1}{b} - 1\right)\left(\frac{1}{c} - 1\right).$$

Solution. Let
$$\sigma_1 = a + b + c, \quad \sigma_2 = ab + bc + ca, \quad \sigma_3 = abc.$$

We will express everything in terms of $\sigma_1, \sigma_2, \sigma_3$. The hypothesis can be written $\sigma_1 = 1$ and $\sigma_2 = 2\sigma_3$. Next, we observe that
$$\left(\frac{1}{a} - 1\right)\left(\frac{1}{b} - 1\right)\left(\frac{1}{c} - 1\right) = \frac{(1-a)(1-b)(1-c)}{abc}$$

and
$$(1-a)(1-b)(1-c) = 1 - \sigma_1 + \sigma_2 - \sigma_3 = \sigma_2 - \sigma_3 = \sigma_3,$$

the last two equalities using the relations $\sigma_1 = 1$ and $\sigma_2 = 2\sigma_3$. We conclude that
$$\left(\frac{1}{a} - 1\right)\left(\frac{1}{b} - 1\right)\left(\frac{1}{c} - 1\right) = \frac{\sigma_3}{\sigma_3} = 1,$$

thus the answer is 1.

Example 11.11. Let a, b, c be real numbers such that $a + b + c = 0$. Prove that
$$\frac{a^7 + b^7 + c^7}{7} = \frac{a^2 + b^2 + c^2}{2} \cdot \frac{a^5 + b^5 + c^5}{5}.$$

Solution. Let $x^3 + Ax^2 + Bx + C$ be the polynomial with roots a, b, c. By Vieta's formulae we have $-A = a+b+c = 0$, thus $A = 0$. Let $S_n = a^n + b^n + c^n$. Then using the results of exercise 1, we compute

$$S_1 = 0, \quad S_2 = (a+b+c)^2 - 2(ab+bc+ca) = -2B$$

and

$$S_3 = 3abc + (a+b+c)((a+b+c)^2 - 3(ab+bc+ca)) = 3abc = -3C.$$

Next, theorem 11.3 gives the recurrence relation

$$S_{n+3} + BS_{n+1} + CS_n = 0$$

for all n, which allows us to compute

$$S_4 = 2B^2, \quad S_5 = 5BC, \quad S_7 = -BS_5 - CS_4 = -7B^2C.$$

We need to prove that

$$\frac{S_7}{7} = \frac{S_2}{2} \cdot \frac{S_5}{5}.$$

But using the previous relations, this comes down to

$$-B^2C = -B \cdot BC,$$

which is clear. The result follows.

Example 11.12. Let $x, y,$ and z be real numbers satisfying $x + y + z = xyz$. Prove that

$$x(1-y^2)(1-z^2) + y(1-z^2)(1-x^2) + z(1-x^2)(1-y^2) = 4xyz.$$

Solution. We try to express the left hand-side in terms of

$$\sigma_1 = x + y + z, \quad \sigma_2 = xy + yz + zx, \sigma_3 = xyz.$$

Note that the hypothesis can be written as $\sigma_1 = \sigma_3$. On the other hand, expanding brutally and rearranging terms in the left hand-side yields

$$x(1-y^2)(1-z^2) + y(1-z^2)(1-x^2) + z(1-x^2)(1-y^2) =$$
$$x - x(y^2+z^2) + xy^2z^2 + y - y(z^2+x^2) + yz^2x^2 + z - z(x^2+y^2) + zx^2y^2 =$$
$$x + y + z + xyz(xy+yz+zx) - x(y^2+z^2) + y(z^2+x^2) + z(x^2+y^2) =$$
$$\sigma_1 + \sigma_1\sigma_2 - x^2(y+z) - y^2(z+x) - z^2(x+y).$$

By exercise 1 we know that

$$x^2(y+z)+y^2(z+x)+z^2(x+y) = (x+y+z)(xy+yz+zx)-3xyz = \sigma_1\sigma_2-3\sigma_3.$$

Hence

$$x(1-y^2)(1-z^2) + y(1-z^2)(1-x^2) + z(1-x^2)(1-y^2) = \sigma_1 + 3\sigma_3 = 4\sigma_3,$$

the last equality being a consequence of the equality $\sigma_1 = \sigma_3$. The result follows.

Example 11.13. Let a, b, c, d be positive integers such that the equation

$$x^2 - (a^2 + b^2 + c^2 + d^2 + 1)x + ab + bc + cd + da = 0$$

has an integer solution. Prove that the other solution is an integer too and both solutions are perfect squares.

Solution. Let x_1, x_2 be the solutions of the equation. Vieta's formulae yield

$$x_1 + x_2 = a^2 + b^2 + c^2 + d^2 + 1, \quad x_1 x_2 = ab + bc + cd + da.$$

The first relation and the fact that one of x_1, x_2 is an integer show that both x_1, x_2 are integers. On the other hand, combining the two relations yields

$$x_1 + x_2 - x_1 x_2 - 1 = a^2 + b^2 + c^2 + d^2 - ab - bc - cd - da =$$

$$\frac{(a-b)^2 + (b-c)^2 + (c-d)^2 + (d-a)^2}{2}.$$

The left hand-side factors as $-(x_1 - 1)(x_2 - 1)$. We deduce that $(x_1 - 1)(x_2 - 1) \leq 0$. But since $x_1 + x_2$ and $x_1 x_2$ are both positive, it follows that so are x_1, x_2. Hence the inequality $(x_1 - 1)(x_2 - 1) \leq 0$ can only happen if $x_1 = 1$ or $x_2 = 1$. Assume without loss of generality that $x_1 = 1$. Then the previous relations yield

$$(a-b)^2 + (b-c)^2 + (c-d)^2 + (d-a)^2 = 0,$$

thus $a = b = c = d$. Coming back to Vieta's relations, we obtain $x_2 = 4a^2 = (2a)^2$, a perfect square. The result follows.

Example 11.14. The equation $x^4 - 18x^3 + kx^2 + 200x - 1984 = 0$ has two solutions whose product is -32. Find k.

Solution. Let x_1, x_2, x_3, x_4 be the solutions of the equation and suppose without loss of generality that $x_1 x_2 = -32$. Since $x_1 x_2 x_3 x_4 = -1984$, we obtain $x_3 x_4 = 62$. Next, we have $x_1 + x_2 + x_3 + x_4 = 18$ and

$$x_2 x_3 x_4 + x_1 x_3 x_4 + x_1 x_2 x_4 + x_1 x_2 x_3 = -200.$$

This last relation combined with $x_1 x_2 = -32$ and $x_3 x_4 = 62$ yields

$$62(x_1 + x_2) - 32(x_3 + x_4) = -200 \quad \text{or} \quad 31(x_1 + x_2) - 16(x_3 + x_4) = -100.$$

Combined with $x_1 + x_2 + x_3 + x_4 = 18$, this gives $x_1 + x_2 = 4$ and $x_3 + x_4 = 14$. We conclude that

$$k = x_1 x_2 + x_2 x_3 + x_3 x_4 + x_4 x_1 + x_1 x_3 + x_2 x_4 =$$

$$-32 + x_2 x_3 + 62 + x_1 x_4 + x_1 x_3 + x_2 x_4 = 30 + (x_1 + x_2)(x_3 + x_4) = 30 + 4 \cdot 14 = 86.$$

Conversely, for this value of k the equation can be written as

$$(x^2 - 4x - 32)(x^2 - 14x + 62) = 0$$

by the previous computations, so it does have two solutions whose product is -32, namely the two solutions of the equation $x^2 - 4x - 32 = 0$.

Example 11.15. Solve in real numbers the system

$$\begin{cases} x + y + z = 0 \\ x^4 + y^4 + z^4 = 18 \\ x^5 + y^5 + z^5 = 30. \end{cases}$$

Solution. Let $a = xy + yz + zx$ and $b = xyz$, so x, y, z are solutions of the equation

$$t^3 + at - b = 0.$$

Let $S_n = x^n + y^n + z^n$. Then theorem 11.3 shows that for all $n \geq 0$

$$S_{n+3} + a S_{n+1} - b S_n = 0.$$

Since $S_1 = 0$, we obtain

$$18 = S_4 = -a S_2 = -a(S_1^2 - 2a) = 2a^2,$$

thus $a^2 = 9$ and $a = \pm 3$. But if $a = 3$, then

$$x^2 + y^2 + z^2 = (x + y + z)^2 - 2(xy + yz + zx) = -2a = -6 < 0,$$

a contradiction. Hence $a = -3$. Next,
$$30 = S_5 = -aS_3 + bS_2 = -3ab - 2ab = -5ab.$$
Thus $ab = -6$ and so $b = 2$, meaning that x, y, z are solutions of the equation $t^3 - 3t - 2 = 0$. This factors as $(t+1)^2(t-2) = 0$ and so its solutions are -1 (with multiplicity 2) and 2. We conclude that the solutions of the system are $(-1, -1, 2)$, $(-1, 2, -1)$ and $(2, -1, -1)$.

Example 11.16. Let A, B be real numbers and let a, b, c be the solutions of the equation $x^3 + Ax + B = 0$. Prove that
$$(a-b)^2(b-c)^2(c-a)^2 = -(4A^3 + 27B^2).$$

Solution. Recall that from Vieta's formulae we have
$$a + b + c = 0, \ ab + bc + ca = A, \ \text{and} \ abc = -B.$$
Note that
$$(a-b)(a-c) = a^2 - ab - ac + bc = 3a^2 + A - 2a(a+b+c) = 3a^2 + A.$$
Symmetrically $(b-c)(b-a) = 3b^2 + A$ and $(c-a)(c-b) = 3c^2 + A$. Hence
$$(a-b)^2(b-c)^2(c-a)^2 = -(3a^2 + A)(3b^2 + B)(3c^2 + C)$$
$$= -27a^2b^2c^2 - 9A(a^2b^2 + b^2c^2 + c^2a^2) - 3A^2(a^2 + b^2 + c^2) - A^3.$$
The coefficients are easy to identify $a^2b^2c^2 = B^2$,
$$a^2b^2 + b^2c^2 + c^2a^2 = (ab + bc + ca)^2 - 2abc(a+b+c) = A^2$$
and
$$a^2 + b^2 + c^2 = (a+b+c)^2 - 2(ab+bc+ca) = -2A.$$
Plugging these in we get
$$(a-b)^2(b-c)^2(c-a)^2 = -27B^2 - 9A^3 + 6A^3 - A^3 = -27B^2 - 4A^3.$$

If you found the first step of the proof tricky, here is an alternative argument, which is certainly longer and more technical, but rather straightforward. Denote for simplicity
$$S = a^2 + b^2 + c^2 = (a+b+c)^2 - 2(ab+bc+ca) = -2A,$$
where the last relation follows from Vieta's formulae
$$a + b + c = 0, \quad ab + bc + ca = A, \quad abc = -B.$$

Note that
$$(a-b)^2(b-c)^2(c-a)^2 = (a^2+b^2-2ab)(b^2+c^2-2bc)(c^2+a^2-2ca) =$$
$$(S-(c^2+2ab))(S-(a^2+2bc))(S-(b^2+2ca)).$$

Expanding brutally the last product gives
$$(a-b)^2(b-c)^2(c-a)^2 = S^3 - S^2(a^2+b^2+c^2+2(ab+bc+ca))+$$
$$S\left[(a^2+2bc)(b^2+2ca)+(b^2+2ca)(c^2+2ab)+(c^2+2ab)(a^2+2bc)\right]-$$
$$(a^2+2bc)(b^2+2ca)(c^2+2ab).$$

Note that
$$a^2+b^2+c^2+2(ab+bc+ca) = (a+b+c)^2 = 0.$$

Next, we deal with
$$(a^2+2bc)(b^2+2ca)+(b^2+2ca)(c^2+2ab)+(c^2+2ab)(a^2+2bc) =$$
$$a^2b^2+b^2c^2+c^2a^2+2ab(a^2+b^2)+2bc(b^2+c^2)+2ca(c^2+a^2)+4abc(a+b+c) =$$
$$a^2b^2+b^2c^2+c^2a^2+2(ab+bc+ca)(a^2+b^2+c^2),$$
the last equality being a consequence of the fact that $a+b+c=0$. Moreover, we have
$$(ab+bc+ca)^2 = a^2b^2+b^2c^2+c^2a^2+2abc(a+b+c) = a^2b^2+b^2c^2+c^2a^2.$$

We conclude that
$$(a^2+2bc)(b^2+2ca)+(b^2+2ca)(c^2+2ab)+(c^2+2ab)(a^2+2bc) = A^2+2AS = -3A^2,$$
since we saw that $S = -2A$.

Finally, another brutal expansion yields
$$(a^2+2bc)(b^2+2ca)(c^2+2ab) = 9(abc)^2+2((ab)^3+(bc)^3+(ca)^3)+4abc(a^3+b^3+c^3).$$

But exercise 1 gives
$$(ab)^3+(bc)^3+(ca)^3 = 3(abc)^2+$$
$$(ab+bc+ca)((ab+bc+ca)^2-3abc(a+b+c)) = 3B^2+A^3$$
and
$$a^3+b^3+c^3 = 3abc+(a+b+c)((a+b+c)^2-3(ab+bc+ca)) = 3B,$$

hence
$$(a^2 + 2bc)(b^2 + 2ca)(c^2 + 2ab) = 2A^3 + 27B^2.$$

Putting everything together yields
$$(a-b)^2(b-c)^2(c-a)^2 = S^3 - 3A^2S - (2A^3 + 27B^2) =$$
$$-8A^3 + 6A^3 - 2A^3 - 27B^2 = -(4A^3 + 27B^2).$$

Example 11.17. Express $(a-b)^2(b-c)^2(c-a)^2$ as a polynomial expression in $a+b+c$, $ab+bc+ca$ and abc.

Solution. We will try to reduce this problem to problem 11.16. Let
$$a' = a - \frac{a+b+c}{3}, \quad b' = b - \frac{a+b+c}{3}, \quad c' = c - \frac{a+b+c}{3}.$$

Then $a' + b' + c' = 0$, hence a', b', c' are roots of some cubic polynomial $x^3 + Ax + B$. Note that
$$(a-b)^2(b-c)^2(c-a)^2 = (a'-b')^2(b'-c')^2(c'-a')^2 = -(4A^3 + 27B^2)$$

by the previous exercise. It remains to express A and B in terms of
$$\sigma_1 = a+b+c, \quad \sigma_2 = ab+bc+ca, \quad \sigma_3 = abc.$$

But a, b, c are roots of the polynomial
$$\left(x - \frac{\sigma_1}{3}\right)^3 + A\left(x - \frac{\sigma_1}{3}\right) + B = x^3 - \sigma_1 x^2 + \left(\frac{\sigma_1^2}{3} + A\right)x + B - \frac{9A\sigma_1 + \sigma_1^3}{27},$$

where $\sigma_1 = a + b + c$. Vieta's relations give
$$\sigma_2 = \frac{\sigma_1^2}{3} + A, \quad \sigma_3 = \frac{9A\sigma_1 + \sigma_1^3}{27} - B,$$

thus finally
$$A = \sigma_2 - \frac{\sigma_1^2}{3}, \quad B = \frac{\sigma_1 \sigma_2}{3} - \frac{2\sigma_1^3}{27} - \sigma_3.$$

Example 11.18. Solve in real numbers the system
$$\begin{cases} a+b+c = 3 \\ a^2 + b^2 + c^2 = 5 \\ (a-b)(b-c)(c-a) = -2. \end{cases}$$

Solution. The first two equations yield (with the usual notations for now)

$$\sigma_1 = 3, \quad \sigma_2 = \frac{\sigma_1^2 - 5}{2} = 2.$$

Using the notations in the solution of the previous problem, we obtain

$$A = \sigma_2 - \frac{\sigma_1^2}{3} = -1, \quad B = \frac{\sigma_1 \sigma_2}{3} - \frac{2\sigma_1^3}{27} - \sigma_3 = -\sigma_3.$$

Hence the previous problem combined with the third equation give

$$4 = (a-b)^2(b-c)^2(c-a)^2 = -(4A^3 + 27B^2) = 4 - 27\sigma_3^2.$$

Thus $\sigma_3 = 0$ and a, b, c are the solutions of the equation

$$t^3 - 3t^2 + 2t = 0 \quad \text{or} \quad t(t-1)(t-2) = 0.$$

We deduce that (a, b, c) is a permutation of $(0, 1, 2)$. Finally, checking the last equation yields the solutions

$$(a, b, c) = (0, 2, 1), \quad (1, 0, 2), \quad (2, 1, 0).$$

12 Introductory problems

1. Al and Bo work on subtracting fractions. Al always subtracts fractions correctly, while Bo does so incorrectly, thinking the rule is
$$\frac{x}{y} - \frac{z}{w} = \frac{x-z}{y+w}.$$
   For some positive integers a, b, c, d, when calculating $\frac{a}{b} - \frac{c}{d}$, both get the same result. Find what Al would obtain if he had to evaluate $\frac{a}{b^2} - \frac{c}{d^2}$.

2. Find all real numbers x such that
$$\sqrt[3]{20 + x\sqrt{2}} + \sqrt[3]{20 - x\sqrt{2}} = 4.$$

3. Prove that for all real numbers a, b, c
$$ab(a-c)(c-b) + bc(b-a)(a-c) + ca(c-b)(b-a) \le 0.$$

4. Simplify the expression
$$\frac{x^3 - 3x + (x^2-1)\sqrt{x^2-4} - 2}{x^3 - 3x + (x^2-1)\sqrt{x^2-4} + 2}$$
   when $x \ge 2$.

5. Prove that
$$(n^3 - 3n + 2)(n^3 + 3n^2 - 4) \quad \text{and} \quad (n^3 - 3n - 2)(n^3 - 3n^2 + 4)$$
   are perfect cubes for all integers n.

6. Let a and b be positive real numbers. Prove that
$$(a+b)^5 \ge 12ab(a^3 + b^3).$$

7. Prove that
$$\sum_{n=1}^{9999} \frac{1}{(\sqrt{n} + \sqrt{n+1})(\sqrt[4]{n} + \sqrt[4]{n+1})} = 9.$$

8. Solve in real numbers the system
$$\begin{cases} a^3 + 3ab^2 + 3ac^2 - 6abc = 1 \\ b^3 + 3ba^2 + 3bc^2 - 6abc = 1 \\ c^3 + 3cb^2 + 3ca^2 - 6abc = 1. \end{cases}$$

9. Real numbers $x_1, x_2, ..., x_{2011}$ are such that $\frac{x_1+x_2}{2}, \frac{x_2+x_3}{2}, ..., \frac{x_{2011}+x_1}{2}$ are a permutation of $x_1, x_2, ..., x_{2011}$. Prove that $x_1 = x_2 = ... = x_{2011}$.

10. Prove that for all $x, y, z \in (-1, 1)$
$$\frac{1}{(1-x)(1-y)(1-z)} + \frac{1}{(1+x)(1+y)(1+z)} \geq 2.$$

11. Find all real numbers x such that
$$(x^2 - x - 2)^4 + (2x+1)^4 = (x^2 + x - 1)^4.$$

12. Let
$$a_n = \sqrt{1 + \left(1 - \frac{1}{n}\right)^2} + \sqrt{1 + \left(1 + \frac{1}{n}\right)^2}.$$
Evaluate $\frac{1}{a_1} + \frac{1}{a_2} + ... + \frac{1}{a_{20}}$.

13. Let $f(x) = ax^2 + bx + c$, where a, b, c are real numbers such that $a > 0$ and $ab \geq \frac{1}{8}$. Prove that $f(b^2 - 4ac) \geq 0$.

14. Factor
$$(x^2 - yz)(y^2 - zx) + (y^2 - zx)(z^2 - xy) + (z^2 - xy)(x^2 - yz).$$

15. Let a, b, c be positive real numbers such that
$$\left(1 + \frac{a}{b}\right)\left(1 + \frac{b}{c}\right)\left(1 + \frac{c}{a}\right) = 9.$$
Prove that
$$\frac{1}{a} + \frac{1}{b} + \frac{1}{c} = \frac{10}{a+b+c}.$$

16. Find the maximal value of
$$\frac{(x+y)(1-xy)}{(1+x^2)(1+y^2)},$$
when $x, y \in \mathbf{R}$.

17. Let a, b be rational numbers such that
$$|a| \leq \frac{47}{|a^2 - 3b^2|} \quad \text{and} \quad |b| \leq \frac{52}{|b^2 - 3a^2|}.$$
Prove that $a^2 + b^2 \leq 17$.

Introductory problems

18. Prove that
$$(1^4 + 1^2 + 1)(2^4 + 2^2 + 1)...(n^4 + n^2 + 1)$$
is not a perfect square if n is a positive integer.

19. Let x, y be real numbers such that
$$(x + \sqrt{x^2 + 1})(y + \sqrt{y^2 + 1}) = 1.$$
Prove that $x + y = 0$.

20. Prove that for all $x, y, z \geq 0$
$$x^2 + xy^2 + xyz^2 \geq 4xyz - 4.$$

21. Let x, y, a be real numbers such that
$$x + y = x^3 + y^3 = x^5 + y^5 = a.$$
Find all possible values of a.

22. Let a and n be positive integers with $n > 1$. Find the integer part of the number
$$(\sqrt[n]{a} - \sqrt[n]{a+1} + \sqrt[n]{a+2})^n.$$

23. Let $f(x) = (x^2 + x)(x^2 + x + 1) + \frac{x^5}{5}$. Evaluate $f(\sqrt[5]{5} - 1)$.

24. Let a, b, c be positive real numbers such that
$$\frac{1}{a} + \frac{1}{b} + \frac{1}{c} = 1.$$
Prove that
$$\frac{1}{a + bc} + \frac{1}{b + ac} + \frac{1}{c + ab} = \frac{2}{(1 + \frac{a}{b})(1 + \frac{b}{c})(1 + \frac{c}{a})}$$

25. Solve the equation
$$\frac{8}{\{x\}} = \frac{9}{x} + \frac{10}{\lfloor x \rfloor}.$$

26. Let a, b, c be positive real numbers not exceeding 1. Prove that
$$(a + b + c)(abc + 1) \geq ab + bc + ca + 3abc.$$

27. Solve in real numbers the system of equations
$$\begin{cases} (x-2y)(x-4z) = 3 \\ (y-2z)(y-4x) = 5 \\ (z-2x)(z-4y) = -8. \end{cases}$$

28. If a, b, c, d are real numbers such that $a^2+b^2+(a+b)^2 = c^2+d^2+(c+d)^2$, prove that
$$a^4 + b^4 + (a+b)^4 = c^4 + d^4 + (c+d)^4.$$

29. The polynomial $P(x) = x^3 + ax^2 + bx + c$ has three distinct real roots. The polynomial $P(Q(x))$, where $Q(x) = x^2+x+2001$, has no real roots. Prove that $P(2001) > \frac{1}{64}$.

30. A sequence $(a_n)_{n \geq 0}$ is defined by $a_0 = 1$ and
$$a_{n+1} = a_{\lfloor \frac{7n}{9} \rfloor} + a_{\lfloor \frac{n}{9} \rfloor}$$
for $n \geq 0$. Prove that there is an n such that $a_n < \frac{n}{2001!}$.

31. Prove that if real numbers a, b satisfy
$$a^2(2a^2 - 4ab + 3b^2) = 3 \quad \text{and} \quad b^2(3a^2 - 4ab + 2b^2) = 5,$$
then $a^3 + b^3 = 2(a+b)$.

32. If x and y are real numbers such that $\frac{x^2+y^2}{x^2-y^2} + \frac{x^2-y^2}{x^2+y^2} = k$, express
$$\frac{x^8 + y^8}{x^8 - y^8} + \frac{x^8 - y^8}{x^8 + y^8}$$
in terms of k.

33. Find all positive real numbers x, y, z satisfying
$$x + \frac{y}{z} = y + \frac{z}{x} = z + \frac{x}{y} = 2.$$

34. Monic quadratic polynomials $f(x)$ and $g(x)$ are such that the equations $f(g(x)) = 0$ and $g(f(x)) = 0$ have no real solutions. Prove that at least one of equations $f(f(x)) = 0$ and $g(g(x)) = 0$ has no real solution (recall that a polynomial is monic if the leading coefficient equals 1).

35. Prove that for all $x, y > 0$
$$\frac{2xy}{x+y} + \sqrt{\frac{x^2+y^2}{2}} \geq \frac{x+y}{2} + \sqrt{xy}.$$

Introductory problems

36. A quadratic polynomial P has the property that $P(x^3+x) \geq P(x^2+1)$ for all real numbers x. Find the sum of the roots of P.

37. Solve the equation
$$x^2 - 10x + 1 = \sqrt{x}(x+1).$$

38. The polynomial $x^3 + ax^2 + bx + c$ has three real roots. Prove that if $-2 \leq a+b+c \leq 0$, then at least one of these roots belongs to $[0, 2]$.

39. Consider the sytem of equations
$$\begin{cases} a_{11}x_1 + a_{12}x_2 + a_{13}x_3 = 0 \\ a_{21}x_1 + a_{22}x_2 + a_{23}x_3 = 0. \\ a_{31}x_1 + a_{32}x_2 + a_{33}x_3 = 0 \end{cases}$$

with unknowns x_1, x_2, x_3. Suppose that:

a) a_{11}, a_{22}, a_{33} are positive numbers;

b) the remaining coefficients are negative numbers;

c) in each equation, the sum of the coefficients is positive.

Prove that the given system has only the solution $x_1 = x_2 = x_3 = 0$.

40. Find all triples (x, y, z) of positive real numbers for which there is a positive real number t such that the following inequalities hold simultaneously:
$$\frac{1}{x} + \frac{1}{y} + \frac{1}{z} + t \leq 4, \quad x^2 + y^2 + z^2 + \frac{2}{t} \leq 5.$$

41. Prove that if $a+b+c = 0$, then
$$\left(\frac{a}{b-c} + \frac{b}{c-a} + \frac{c}{a-b}\right)\left(\frac{b-c}{a} + \frac{c-a}{b} + \frac{a-b}{c}\right) = 9.$$

42. Let a, b, c, x, y, z, m, n be positive real numbers satisfying
$$\sqrt[3]{a} + \sqrt[3]{b} + \sqrt[3]{c} = \sqrt[3]{m}, \quad \sqrt{x} + \sqrt{y} + \sqrt{z} = \sqrt{n}.$$

Prove that
$$\frac{a}{x} + \frac{b}{y} + \frac{c}{z} \geq \frac{m}{n}.$$

43. Find all triples x, y, z of positive integers such that
$$x^2y + y^2z + z^2x = xy^2 + yz^2 + zx^2 = 111.$$

44. Let a, b, c be nonzero real numbers such that
$$\frac{1}{a^3} + \frac{1}{b^3} + \frac{1}{c^3} = \frac{1}{a^3 + b^3 + c^3}.$$
Prove that
$$\frac{1}{a^5} + \frac{1}{b^5} + \frac{1}{c^5} = \frac{1}{a^5 + b^5 + c^5}.$$

45. Solve in nonzero real numbers the system of equations
$$\begin{cases} \frac{7}{x} - \frac{27}{y} = 2x^2 \\ \frac{9}{y} - \frac{21}{x} = 2y^2. \end{cases}$$

46. Compute the integer part of the number
$$S = \sqrt{2} + \sqrt[3]{\frac{3}{2}} + \ldots + \sqrt[2013]{\frac{2013}{2012}}.$$

47. Prove that for all $a, b, c > 0$
$$\frac{a^3}{b^2 + c^2} + \frac{b^3}{c^2 + a^2} + \frac{c^3}{a^2 + b^2} \geq \frac{a+b+c}{2}.$$

48. Solve in real numbers the equation $x^3 + 1 = 2\sqrt[3]{2x - 1}$.

49. Prove that for all natural numbers n,
$$\sum_{k=1}^{n^2} \{\sqrt{k}\} \leq \frac{n^2 - 1}{2}.$$

50. Let a, b be real numbers. Solve in real numbers the system
$$\begin{cases} x + y = \sqrt[3]{a + b} \\ x^4 - y^4 = ax - by. \end{cases}$$

51. Prove that for all $a, b, c > 0$
$$\frac{abc}{a^3 + b^3 + abc} + \frac{abc}{b^3 + c^3 + abc} + \frac{abc}{c^3 + a^3 + abc} \leq 1.$$

52. Define a sequence $(a_n)_{n \geq 1}$ by
$$a_1 = 1, \quad a_{n+1} = \frac{1 + 4a_n + \sqrt{1 + 24a_n}}{16}.$$
Find an explicit formula for a_n.

13 Advanced problems

1. If x, y are positive real numbers, define
$$x * y = \frac{x+y}{1+xy}.$$
Find $(...(((2*3)*4)*5)...)*1995$.

2. Three real numbers are given, such that the fractional part of the product of every two of them is $\frac{1}{2}$. Prove that these numbers are irrational.

3. Let a, b, c be real numbers satisfying
$$\begin{cases} (a+b)(b+c)(c+a) = abc \\ (a^3+b^3)(b^3+c^3)(c^3+a^3) = a^3b^3c^3. \end{cases}$$
Prove that $abc = 0$.

4. Real numbers a, b satisfy $a^3 - 3a^2 + 5a - 17 = 0$ and $b^3 - 3b^2 + 5b + 11 = 0$. Find $a + b$.

5. Prove that
$$\sqrt{1 + \frac{1}{1^2} + \frac{1}{2^2}} + \sqrt{1 + \frac{1}{2^2} + \frac{1}{3^2}} + ... + \sqrt{1 + \frac{1}{1999^2} + \frac{1}{2000^2}}$$
is a rational number and compute it in lowest form.

6. Find a polynomial with integer coefficients having $\sqrt[5]{2 + \sqrt{3}} + \sqrt[5]{2 - \sqrt{3}}$ as a root.

7. Solve in real numbers the equation
$$(3x+1)(4x+1)(6x+1)(12x+1) = 5.$$

8. Let n be a positive integer. Solve in real numbers the equation
$$\lfloor x \rfloor + \lfloor 2x \rfloor + ... + \lfloor nx \rfloor = \frac{n(n+1)}{2}.$$

9. Any two of the real numbers a_1, a_2, a_3, a_4, a_5 differ by no less than 1. Moreover, there is a real number k satisfying
$$\begin{cases} a_1 + a_2 + a_3 + a_4 + a_5 = 2k \\ a_1^2 + a_2^2 + a_3^2 + a_4^2 + a_5^2 = 2k^2. \end{cases}$$
Prove that $k^2 \geq \frac{25}{3}$.

10. Prove that if a, b, c, d are nonzero real numbers, not all equal, and if
$$a + \frac{1}{b} = b + \frac{1}{c} = c + \frac{1}{d} = d + \frac{1}{a},$$
then $|abcd| = 1$.

11. The side lengths of a triangle are the roots of a cubic polynomial with rational coefficients. Prove that the altitudes of this triangle are roots of a polynomial of sixth degree with rational coefficients.

12. Find the maximal value of
$$\frac{(1+x)^8 + 16x^4}{(1+x^2)^4}$$
when $x \in \mathbf{R}$.

13. Let x, y, z be real numbers greater than -1. Prove that
$$\frac{1+x^2}{1+y+z^2} + \frac{1+y^2}{1+z+x^2} + \frac{1+z^2}{1+x+y^2} \geq 2.$$

14. Let a and b be real numbers. Prove that
$$a^3 + b^3 + (a+b)^3 + 6ab = 16$$
if and only if $a + b = 2$.

15. If a and b are nonzero real numbers such that
$$20a + 21b = \frac{a}{a^2+b^2} \quad \text{and} \quad 21a - 20b = \frac{b}{a^2+b^2},$$
evaluate $a^2 + b^2$.

16. Let a, b be solutions of the equation $x^4 + x^3 - 1 = 0$. Prove that ab is a solution of the equation $x^6 + x^4 + x^3 - x^2 - 1 = 0$.

17. Find all real numbers x such that
$$\sqrt{x + 2\sqrt{x + 2\sqrt{x + 2\sqrt{3x}}}} = x.$$

18. Let a_1, a_2, a_3, a_4, a_5 be real numbers satisfying
$$\frac{a_1}{k^2+1} + \frac{a_2}{k^2+2} + \frac{a_3}{k^2+3} + \frac{a_4}{k^2+4} + \frac{a_5}{k^2+5} = \frac{1}{k^2}$$
for $1 \leq k \leq 5$. Find the value of $\frac{a_1}{37} + \frac{a_2}{38} + \frac{a_3}{39} + \frac{a_4}{40} + \frac{a_5}{41}$.

Advanced problems

19. Are there nonzero real numbers a, b, c such that for all $n > 3$ there is a polynomial $P_n(x) = x^n + \cdots + ax^2 + bx + c$ which has exactly n (not necessarily distinct) integral roots?

20. Let $n > 1$ be an integer and let $a_0, a_1, ..., a_n$ be real numbers such that $a_0 = \frac{1}{2}$ and
$$a_{k+1} = a_k + \frac{a_k^2}{n} \quad \text{for} \quad k = 0, 1, ...n-1.$$
Prove that $1 - \frac{1}{n} < a_n < 1$.

21. Let $x_i = \frac{i}{101}$. Compute
$$\sum_{i=0}^{101} \frac{x_i^3}{1 - 3x_i + 3x_i^2}.$$

22. Let a, b be real numbers and let $f(x) = x^2 + ax + b$. Suppose that the equation $f(f(x)) = 0$ has four different real solutions, and that the sum of two of these solutions is -1. Prove that $b \leq -\frac{1}{4}$.

23. Real numbers a, b, c satisfy
$$\frac{a}{a^2 - bc} + \frac{b}{b^2 - ca} + \frac{c}{c^2 - ab} = 0.$$
Prove that
$$\frac{a}{(a^2 - bc)^2} + \frac{b}{(b^2 - ca)^2} + \frac{c}{(c^2 - ab)^2} = 0.$$

24. Let n be a positive integer and let $a_k = 2^{2^{k-n}} + k$. Prove that
$$(a_1 - a_0)(a_2 - a_1)...(a_n - a_{n-1}) = \frac{7}{a_0 + a_1}.$$

25. Find the integer part of
$$1 + \frac{1}{\sqrt[3]{2^2}} + \frac{1}{\sqrt[3]{3^2}} + ... + \frac{1}{\sqrt[3]{(10^9)^2}}.$$

26. Is there a sequence $(a_n)_{n \geq 1}$ of positive real numbers such that
$$a_1 + a_2 + ... + a_n \leq n^2 \quad \text{and} \quad \frac{1}{a_1} + \frac{1}{a_2} + ... + \frac{1}{a_n} \leq 2008$$
for all positive integers n?

27. Is there a polynomial f with integer coefficients such that $f(x,y,z)$ and $x + \sqrt[3]{2}y + \sqrt[3]{3}z$ have the same sign for all integers x, y, z?

28. Prove that there are infinitely many odd numbers in the sequence $\lfloor n\sqrt{2} \rfloor + \lfloor n\sqrt{3} \rfloor$ (with $n \geq 1$).

29. Prove that if $x_1, ..., x_n > 0$ satisfy $x_1 x_2 ... x_n = 1$, then
$$\left(\frac{x_1 + x_2 + ... + x_n}{n} \right)^{2n} \geq \frac{x_1^2 + x_2^2 + ... + x_n^2}{n}.$$

30. The equation $x^3 + x^2 - 2x - 1 = 0$ has three real solutions x_1, x_2, x_3. Find $\sqrt[3]{x_1} + \sqrt[3]{x_2} + \sqrt[3]{x_3}$.

31. Prove that for all $a, b, c, x, y, z \geq 0$
$$(a^2 + x^2)(b^2 + y^2)(c^2 + z^2) \geq (ayz + bzx + cxy - xyz)^2.$$

32. The sequence a_n is defined by $a_1 = \frac{1}{2}$ and
$$a_{n+1} = \frac{a_n^2}{a_n^2 - a_n + 1}$$
for $n \geq 1$. Prove that $a_1 + a_2 + ... + a_n < 1$ for all positive integers n.

33. Prove that for all $a, b, c > 0$
$$\frac{a+b+c}{\sqrt[3]{abc}} + \frac{8abc}{(a+b)(b+c)(c+a)} \geq 4.$$

34. Find all real numbers x such that
$$\frac{x^2}{x-1} + \sqrt{x-1} + \frac{\sqrt{x-1}}{x^2} = \frac{x-1}{x^2} + \frac{1}{\sqrt{x-1}} + \frac{x^2}{\sqrt{x-1}}.$$

35. Let $x > 30$ be a real number such that $\lfloor x \rfloor \cdot \lfloor x^2 \rfloor = \lfloor x^3 \rfloor$. Prove that $\{x\} < \frac{1}{2700}$.

36. Prove that for all $x, y > 0$ we have $x^y + y^x > 1$.

37. Solve in real numbers the system
$$\begin{cases} x^3 + x(y-z)^2 = 2 \\ y^3 + y(z-x)^2 = 30 \\ z^3 + z(x-y)^2 = 16. \end{cases}$$

Advanced problems

38. Let a, b, c, d be positive real numbers with $a+b+c+d = 4$. Prove that
$$\frac{a^4}{(a+b)(a^2+b^2)} + \frac{b^4}{(b+c)(b^2+c^2)} + \frac{c^4}{(c+d)(c^2+d^2)} + \frac{d^4}{(d+a)(d^2+a^2)}$$
is equal to at least 1.

39. Nonnegative numbers a, b, c, d, e, f add up to 6. Find the maximal value of
$$abc + bcd + cde + def + efa + fab.$$

40. Solve the following equation
$$\lfloor x \rfloor + \lfloor 2x \rfloor + \lfloor 4x \rfloor + \lfloor 8x \rfloor + \lfloor 16x \rfloor + \lfloor 32x \rfloor = 12345.$$

41. Let x, y, z be real numbers such that $x + y + z = 0$. Prove that
$$\frac{x(x+2)}{2x^2+1} + \frac{y(y+2)}{2y^2+1} + \frac{z(z+2)}{2z^2+1} \geq 0.$$

42. Solve in real numbers the system
$$\begin{cases} \frac{1}{x} + \frac{1}{2y} = (x^2+3y^2)(3x^2+y^2) \\ \frac{1}{x} - \frac{1}{2y} = 2(y^4-x^4). \end{cases}$$

43. Find all triples (x, y, z) of positive real numbers satisfying simultaneously the inequalities $x+y+z-2xyz \leq 1$ and
$$xy + yz + zx + \frac{1}{xyz} \leq 4.$$

44. Let a, b, c be positive real numbers and let $x = a + \frac{1}{b}$, $y = b + \frac{1}{c}$, $z = c + \frac{1}{a}$. Prove that
$$xy + yz + zx \geq 2(x+y+z).$$

45. Let a, b be nonzero real numbers, such that $\lfloor an + b \rfloor$ is an even integer for all positive integers n. Prove that a is an even integer.

46. Let a, b, c be positive real numbers. Find all real numbers x, y, z such that
$$\begin{cases} ax + by = (x-y)^2 \\ by + cz = (y-z)^2 \\ cz + ax = (z-x)^2. \end{cases}$$

47. Positive real numbers a, b, c add up to 1. Prove that
$$(ab + bc + ca)\left(\frac{a}{b^2 + b} + \frac{b}{c^2 + c} + \frac{c}{a^2 + a}\right) \geq \frac{3}{4}.$$

48. A sequence of nonnegative real numbers $(a_n)_{n \geq 1}$ satisfies
$$|a_m - a_n| \geq \frac{1}{m + n}$$
for all distinct positive integers m, n. Prove that if a real number c is greater than a_n for all $n \geq 1$, then $c \geq 1$.

49. Let a, b, c, d be real numbers such that $a + b + c + d = 0$ and $a^7 + b^7 + c^7 + d^7 = 0$. Find all possible values of $a(a+b)(a+c)(a+d)$.

50. Let a, b, c be real numbers such that $a^2 + b^2 + c^2 = 9$. Prove that
$$2(a + b + c) - abc \leq 10.$$

51. A real number $x > 1$ has the property that $x^n\{x^n\} < \frac{1}{4}$ for all $n \geq 1$. Prove that x is an integer.

52. Is there a sequence $(a_n)_{n \geq 1}$ of real numbers such that $a_n \in [0, 4]$ for all n and
$$|a_m - a_n| \geq \frac{1}{|m - n|}$$
for all distinct positive integers m, n?

53. Let $a, b, c, d \in [\frac{1}{2}, 2]$ be such that $abcd = 1$. Find the maximal value of
$$\left(a + \frac{1}{b}\right)\left(b + \frac{1}{c}\right)\left(c + \frac{1}{d}\right)\left(d + \frac{1}{a}\right).$$

14 Solutions to introductory problems

1. Al and Bo work on subtracting fractions. Al always subtracts fractions correctly, while Bo does so incorrectly, thinking the rule is

$$\frac{x}{y} - \frac{z}{w} = \frac{x-z}{y+w}.$$

   For some positive integers a, b, c, d, when calculating $\frac{a}{b} - \frac{c}{d}$, both get the same result. Find what Al would obtain if he had to evaluate $\frac{a}{b^2} - \frac{c}{d^2}$.

   Solution. Since Al and Bo obtained the same result, we must have

$$\frac{ad - bc}{bd} = \frac{a}{b} - \frac{c}{d} = \frac{a-c}{b+d}.$$

   This can also be written

$$(ad - bc)(b + d) = (a - c)bd.$$

   Expanding gives

$$abd + ad^2 - b^2 c - bcd = abd - bcd,$$

   which simplifies to $ad^2 - b^2 c = 0$. Hence Al would obtain

$$\frac{a}{b^2} - \frac{c}{d^2} = \frac{ad^2 - b^2 c}{b^2 d^2} = 0.$$

2. Find all real numbers x such that

$$\sqrt[3]{20 + x\sqrt{2}} + \sqrt[3]{20 - x\sqrt{2}} = 4.$$

   Solution. Let us take the cube of the original relation and use the identity

$$(a + b)^3 = a^3 + b^3 + 3ab(a + b).$$

   We obtain

$$64 = 40 + 3\sqrt[3]{400 - 2x^2} \cdot 4$$

   and then $x^2 = 196$, with the solutions $x = \pm 14$. We conclude that the equation has two solutions, namely 14 and -14.

   Let us give an alternative approach: we introduce two new variables $a = \sqrt[3]{20 + x\sqrt{2}}$ and $b = \sqrt[3]{20 - x\sqrt{2}}$. The equation becomes $a + b = 4$, but there is a hidden relationship between a and b. Indeed, taking the cube of a and b we obtain

$$a^3 = 20 + x\sqrt{2}, \quad b^3 = 20 - x\sqrt{2},$$

hence by adding these two relations $a^3 + b^3 = 40$. On the other hand, since $a + b = 4$, we have

$$a^3 + b^3 = (a+b)(a^2 - ab + b^2) = 4((a+b)^2 - 3ab) = 4(16 - 3ab) = 64 - 12ab,$$

thus
$$ab = \frac{64 - 40}{12} = 2.$$

On the other hand, $ab = \sqrt[3]{400 - 2x^2}$, thus $400 - 2x^2 = 8$ and then $x^2 = 196$ and $x = \pm 14$. It follows that the solutions of the equation are 14 and -14.

3. Prove that for all real numbers a, b, c

$$ab(a-c)(c-b) + bc(b-a)(a-c) + ca(c-b)(b-a) \leq 0.$$

Solution. We expand the left hand-side and rewrite the inequality as

$$ab(ac - ab - c^2 + bc) + bc(ab - bc - a^2 + ca) + ca(bc - ca - b^2 + ab) \leq 0.$$

This is turn simplifies to

$$a^2b^2 + b^2c^2 + c^2a^2 \geq abc(a+b+c),$$

which is equivalent to

$$(ab - bc)^2 + (bc - ca)^2 + (ca - ab)^2 \geq 0$$

and thus true.

4. Simplify
$$\frac{x^3 - 3x + (x^2 - 1)\sqrt{x^2 - 4} - 2}{x^3 - 3x + (x^2 - 1)\sqrt{x^2 - 4} + 2}$$

when $x \geq 2$.

Solution. Let us look separately at the numerator and at the denominator. For the numerator, we will study the term $x^3 - 3x - 2$. It vanishes at $x = -1$ and $x = 2$, thus we can factor

$$x^3 - 3x - 2 = x^3 - x - 2(x+1) = x(x-1)(x+1) - 2(x+1) =$$
$$(x+1)(x^2 - x - 2) = (x+1)(x-2)(x+1) = (x+1)^2(x-2).$$

Thus

$$x^3 - 3x + (x^2 - 1)\sqrt{x^2 - 4} - 2 = (x+1)^2(x-2) + (x-1)(x+1)\sqrt{x^2 - 4} =$$

$$(x+1)\sqrt{x-2}((x+1)\sqrt{x-2}+(x-1)\sqrt{x+2}).$$

We do the same with the denominator, factoring

$$x^3 - 3x + 2 = x^3 - x - 2(x-1) = (x-1)(x^2+x-2) = (x-1)^2(x+2)$$

and

$$x^3 - 3x + (x^2-1)\sqrt{x^2-4} + 2$$
$$= (x-1)\sqrt{x+2}((x-1)\sqrt{x+2}+(x+1)\sqrt{x-2}).$$

We conclude that

$$\frac{x^3 - 3x + (x^2-1)\sqrt{x^2-4} - 2}{x^3 - 3x + (x^2-1)\sqrt{x^2-4} + 2} = \frac{x+1}{x-1} \cdot \sqrt{\frac{x-2}{x+2}}.$$

5. Prove that

$$(n^3 - 3n + 2)(n^3 + 3n^2 - 4) \quad \text{and} \quad (n^3 - 3n - 2)(n^3 - 3n^2 + 4)$$

are perfect cubes for all integers n.

Solution. Let us deal with the first expression first. A key observation is that both $n^3 - 3n + 2$ and $n^3 + 3n^2 - 4$ vanish when $n = 1$. Hence we will be able to factor $n - 1$ out of each of them. Indeed,

$$n^3 - 3n + 2 = n^3 - n - 2n + 2 = n(n^2 - 1) - 2(n-1)$$
$$= (n-1)n(n+1) - 2(n-1) = (n-1)(n^2 + n - 2)$$

and similarly
$$n^3 + 3n^2 - 4 = n^3 - 1 + 3n^2 - 3 =$$
$$(n-1)(n^2 + n + 1) + 3(n-1)(n+1) = (n-1)(n^2 + n + 1 + 3n + 3).$$

The good news is that the expressions n^2+n-2 and $n^2+n+1+3n+3 = n^2+4n+4$ can also be easily factored. Indeed, the first one vanishes at $n = 1$, so as above we end up with $n^2 + n - 2 = (n-1)(n+2)$, while the second one equals $(n+2)^2$. Putting everything together, we obtain

$$(n^3 - 3n + 2)(n^3 + 3n^2 - 4) = (n-1)^2(n+2)(n-1)(n+2)^2$$
$$= (n-1)^3 \cdot (n+2)^3 = [(n-1)(n+2)]^3,$$

which is a perfect cube.

With similar arguments we obtain

$$n^3 - 3n - 2 = n^3 - n - 2(n+1) = n(n-1)(n+1) - 2(n+1)$$

$$= (n^2 - n - 2)(n + 1) = (n + 1)^2(n - 2)$$

and
$$n^3 - 3n^2 + 4 = n^3 + n^2 - 4(n^2 - 1) =$$
$$n^2(n + 1) - 4(n - 1)(n + 1) = (n + 1)(n - 2)^2,$$

hence
$$(n^3 - 3n - 2)(n^3 - 3n^2 + 4) = [(n + 1)(n - 2)]^3,$$

a perfect cube again.

6. Let a and b be positive real numbers. Prove that
$$(a + b)^5 \geq 12ab(a^3 + b^3).$$

Solution. We work with
$$s = a + b \quad \text{and} \quad p = ab.$$

The inequality reduces to
$$s^4 \geq 12p(s^2 - 3p),$$

which is equivalent to
$$(s^2 - 6p)^2 \geq 0$$

and thus true.

7. Prove that
$$\sum_{n=1}^{9999} \frac{1}{(\sqrt{n} + \sqrt{n+1})(\sqrt[4]{n} + \sqrt[4]{n+1})} = 9.$$

Solution. The key point is the equality
$$(\sqrt{a} + \sqrt{b})(\sqrt[4]{a} + \sqrt[4]{b}) = \frac{a - b}{\sqrt[4]{a} - \sqrt[4]{b}},$$

which is a consequence of the equality
$$(x - y)(x + y)(x^2 + y^2) = x^4 - y^4,$$

with $x = \sqrt[4]{a}$ and $y = \sqrt[4]{b}$. Hence
$$\sum_{n=1}^{9999} \frac{1}{(\sqrt{n} + \sqrt{n+1})(\sqrt[4]{n} + \sqrt[4]{n+1})} =$$
$$= \sum_{n=1}^{9999} \left(\sqrt[4]{n+1} - \sqrt[4]{n} \right) = \sqrt[4]{10000} - \sqrt[4]{1} = 10 - 1 = 9.$$

8. Solve in real numbers the system
$$\begin{cases} a^3 + 3ab^2 + 3ac^2 - 6abc = 1 \\ b^3 + 3ba^2 + 3bc^2 - 6abc = 1 \\ c^3 + 3cb^2 + 3ca^2 - 6abc = 1. \end{cases}$$

Solution. Let us subtract the first two equations. We obtain
$$a^3 + 3ab^2 + 3ac^2 = b^3 + 3ba^2 + 3bc^2.$$

This can be factored as follows:
$$a^3 + 3ab^2 + 3ac^2 - b^3 - 3ba^2 - 3bc^2 =$$
$$a^3 - b^3 + 3ab(b-a) + 3c^2(a-b) = (a-b)(a^2 + ab + b^2 - 3ab + 3c^2).$$

We obtain therefore
$$(a-b)((a-b)^2 + 3c^2) = 0.$$

This necessarily implies $a = b$ (since the equality $(a-b)^2 + 3c^2 = 0$ forces $a = b$). Doing the same with the second and third equations, we obtain $a = b = c$, and then the system reduces to $a^3 = 1$. Thus $(1, 1, 1)$ is the unique solution of the problem.

9. Real numbers $x_1, x_2, ..., x_{2011}$ are such that $\frac{x_1+x_2}{2}, \frac{x_2+x_3}{2}, ..., \frac{x_{2011}+x_1}{2}$ are a permutation of $x_1, x_2, ..., x_{2011}$. Prove that $x_1 = x_2 = ... = x_{2011}$.

Solution. Let $n = 2011$. If $y_1, ..., y_n$ is a permutation of $x_1, ..., x_n$, then certainly
$$y_1^2 + y_2^2 + ... + y_n^2 = x_1^2 + ... + x_n^2.$$

In our case, we obtain
$$\frac{(x_1+x_2)^2}{4} + ... + \frac{(x_n+x_1)^2}{4} = x_1^2 + ... + x_n^2,$$

which becomes after expansion
$$x_1 x_2 + ... + x_n x_1 = x_1^2 + ... + x_n^2.$$

This can also be written as
$$(x_1 - x_2)^2 + ... + (x_n - x_1)^2 = 0,$$

forcing $x_1 = ... = x_n$ and solving the problem.

Another approach is based on the fact that

$$\frac{(x_i + x_{i+1})^2}{4} \leq \frac{x_i^2 + x_{i+1}^2}{2}$$

for all $i = 1, ..., n$. Since adding these inequalities yields an equality by hypothesis, it follows that each of them is an equality, which happens precisely when $x_i = x_{i+1}$ for all i, that is $x_1 = ... = x_n$.

10. Prove that for all $x, y, z \in (-1, 1)$ we have

$$\frac{1}{(1-x)(1-y)(1-z)} + \frac{1}{(1+x)(1+y)(1+z)} \geq 2.$$

Solution. There is a very smart and short proof using the AM-GM inequality: by hypothesis, both terms in the sum appearing in the left hand-side are positive, hence using the inequality $a + b \geq 2\sqrt{ab}$ we obtain

$$\frac{1}{(1-x)(1-y)(1-z)} + \frac{1}{(1+x)(1+y)(1+z)}$$
$$\geq 2 \cdot \frac{1}{\sqrt{(1-x^2)(1-y^2)(1-z^2)}}.$$

Thus it suffices to prove that $(1 - x^2)(1 - y^2)(1 - z^2) \leq 1$, which is clear, since $1 - x^2, 1 - y^2, 1 - z^2 \in (0, 1]$. We have equality if and only if $x = y = z = 0$.

11. Find all real numbers x such that

$$(x^2 - x - 2)^4 + (2x + 1)^4 = (x^2 + x - 1)^4.$$

Solution. Let us set $a = x^2 - x - 2$ and $b = 2x + 1$. Then $a + b = x^2 + x - 1$, thus the equation can be written as

$$a^4 + b^4 = (a + b)^4.$$

Expanding
$$(a + b)^4 = a^4 + 4a^3b + 6a^2b^2 + 4ab^3 + b^4,$$

we obtain
$$2ab(2a^2 + 3ab + 2b^2) = 0.$$

If $a = 0$, we obtain $x^2 - x - 2 = 0$, thus $x = -1$ or $x = 2$, which are solutions of the equation. If $b = 0$, then $x = -\frac{1}{2}$, another solution. Finally, suppose that $ab \neq 0$, then the preceding discussion shows that $2a^2 + 3ab + 2b^2 = 0$. Seeing this as a quadratic equation in a, we compute the discriminant $\Delta = -7b^2 < 0$, hence the equation has no solution. We conclude that the original equation has three solutions: $-1, -\frac{1}{2}$ and 2.

12. Let
$$a_n = \sqrt{1 + \left(1 - \frac{1}{n}\right)^2} + \sqrt{1 + \left(1 + \frac{1}{n}\right)^2}.$$
Evaluate $\frac{1}{a_1} + \frac{1}{a_2} + \ldots + \frac{1}{a_{20}}$.

Solution. We have
$$a_n = \sqrt{1 + \frac{n^2 - 2n + 1}{n^2}} + \sqrt{1 + \frac{n^2 + 2n + 1}{n^2}}$$
$$= \frac{1}{n}\left(\sqrt{2n^2 - 2n + 1} + \sqrt{2n^2 + 2n + 1}\right).$$

Using the identity
$$\frac{1}{\sqrt{x} + \sqrt{y}} = \frac{\sqrt{x} - \sqrt{y}}{x - y},$$

we deduce that
$$\frac{1}{a_n} = \frac{\sqrt{2n^2 + 2n + 1} - \sqrt{2n^2 - 2n + 1}}{4}.$$

It follows that
$$\frac{1}{a_1} + \frac{1}{a_2} + \ldots + \frac{1}{a_{20}} = \frac{\sqrt{5} - 1}{4} + \frac{\sqrt{13} - \sqrt{5}}{4} + \ldots + \frac{\sqrt{841} - \sqrt{761}}{4}$$

and the last expression is a telescoping sum. We conclude that
$$\frac{1}{a_1} + \frac{1}{a_2} + \ldots + \frac{1}{a_{20}} = \frac{29 - 1}{4} = 7.$$

13. Let $f(x) = ax^2 + bx + c$, where a, b, c are real numbers such that $a > 0$ and $ab \geq \frac{1}{8}$. Prove that $f(b^2 - 4ac) \geq 0$.

Solution. Let $\Delta = b^2 - 4ac$. If $\Delta \leq 0$, then $f(x) \geq 0$ for all $x \in \mathbf{R}$ (since we assumed that $a > 0$), so $f(\Delta) \geq 0$. If $\Delta > 0$ and $x_1 < x_2$ are the two real roots of the equation $f(x) = 0$, then the inequality $f(\Delta) \geq 0$ is equivalent (since $a > 0$) to
$$(\Delta - x_1)(\Delta - x_2) \geq 0.$$

It suffices therefore to prove that
$$\Delta \geq x_2 = \frac{-b + \sqrt{\Delta}}{2a}.$$

But the AM-GM inequality yields

$$\Delta + \frac{b}{2a} \geq 2\sqrt{\Delta \cdot \frac{b}{2a}}.$$

By hypothesis, the last quantity is greater than or equal to

$$2\sqrt{\Delta \cdot \frac{1}{16a^2}} = \frac{\sqrt{\Delta}}{2a}.$$

Thus

$$\Delta + \frac{b}{2a} \geq \frac{\sqrt{\Delta}}{2a},$$

that is $\Delta \geq x_2$.

14. Factor

$$(x^2 - yz)(y^2 - zx) + (y^2 - zx)(z^2 - xy) + (z^2 - xy)(x^2 - yz).$$

Solution. When no smart idea appears, brute force usually gives at least the first step. So expanding brutally the given expression (call it E) we obtain

$$E = x^2y^2 - x^3z - y^3z + z^2xy$$
$$+ y^2z^2 - y^3x - z^3x + x^2yz + z^2x^2 - z^3y - x^3y + y^2xz.$$

One can hardly see anything but darkness in this mess, but another natural idea is to make it symmetric by collecting similar terms. This yields

$$E = (x^2y^2 + y^2z^2 + z^2x^2) + xyz(x + y + z)$$
$$- xy(x^2 + y^2) - yz(y^2 + z^2) - zx(z^2 + x^2).$$

Now, let us observe that

$$xy(x^2 + y^2) = xy(x^2 + y^2 + z^2) - xyz^2.$$

Writing down similar expressions and adding them up yields

$$xy(x^2 + y^2) + yz(y^2 + z^2) + zx(z^2 + x^2)$$
$$= (xy + yz + zx)(x^2 + y^2 + z^2) - xyz^2 - yzx^2 - zxy^2.$$

Hence

$$E = x^2y^2 + y^2z^2 + z^2x^2 + 2xyz(x + y + z) - (xy + yz + zx)(x^2 + y^2 + z^2).$$

Solutions to introductory problems

And now we are saved, since

$$x^2y^2 + y^2z^2 + z^2x^2 + 2xyz(x+y+z) = (xy)^2 + (yz)^2 + (zx)^2$$
$$+ 2(xy)(yz) + 2(yz)(zx) + 2(zx)(xy) = (xy + yz + zx)^2.$$

Putting everything together yields

$$E = (xy + yz + zx)(xy + yz + zx - x^2 - y^2 - z^2).$$

15. Let a, b, c be positive real numbers such that

$$\left(1 + \frac{a}{b}\right)\left(1 + \frac{b}{c}\right)\left(1 + \frac{c}{a}\right) = 9.$$

Prove that

$$\frac{1}{a} + \frac{1}{b} + \frac{1}{c} = \frac{10}{a+b+c}.$$

Solution. We will work on two different directions, rewriting both the hypothesis and the conclusion in more convenient terms. First, we expand the product to rewrite the hypothesis as follows:

$$9 = \left(1 + \frac{a}{b}\right)\left(1 + \frac{b}{c}\right)\left(1 + \frac{c}{a}\right) = 1 + \frac{a}{b} + \frac{b}{c} + \frac{c}{a} + \frac{a}{b}\cdot\frac{b}{c} + \frac{b}{c}\cdot\frac{c}{a} + \frac{c}{a}\cdot\frac{a}{b} + 1,$$

thus

$$7 = \frac{a}{b} + \frac{b}{c} + \frac{c}{a} + \frac{a}{c} + \frac{b}{a} + \frac{c}{b}.$$

Since there is no apparent way of simplifying this, let us turn to the conclusion of the problem, which we write as

$$10 = (a+b+c)\left(\frac{1}{a} + \frac{1}{b} + \frac{1}{c}\right) = 1 + \frac{a}{b} + \frac{a}{c} + \frac{b}{a} + 1 + \frac{b}{c} + \frac{c}{a} + \frac{c}{b} + 1.$$

Contemplating a few seconds this equality, we see that it is simply the equivalent form of the hypothesis that we have already established.

16. Find the maximal value of

$$\frac{(x+y)(1-xy)}{(1+x^2)(1+y^2)},$$

when $x, y \in \mathbf{R}$.

Solution. The key idea is Lagrange's identity

$$(1+x^2)(1+y^2) = (x+y)^2 + (1-xy)^2.$$

Letting $a = x + y$ and $b = 1 - xy$, we obtain

$$\frac{(x+y)(1-xy)}{(1+x^2)(1+y^2)} = \frac{ab}{a^2+b^2} \leq \frac{1}{2}.$$

It remains to see whether the value $\frac{1}{2}$ is attained (if this is not the case, then what we have already done is not terribly useful...). This happens if and only if $\frac{ab}{a^2+b^2} = \frac{1}{2}$, which can be written as $(a-b)^2 = 0$, or $a = b$. So we need to find real numbers x, y such that $x + y = 1 - xy$. Well, simply choose $y = 0$ and $x = 1$.

17. Let a, b be rational numbers such that

$$|a| \leq \frac{47}{|a^2 - 3b^2|} \quad \text{and} \quad |b| \leq \frac{52}{|b^2 - 3a^2|}.$$

Prove that $a^2 + b^2 \leq 17$.

Solution. We write the hypothesis as

$$|a^3 - 3ab^2| \leq 47 \quad \text{and} \quad |b^3 - 3a^2b| \leq 52.$$

Let

$$x = a^3 - 3ab^2, \quad y = b^3 - 3a^2b.$$

Then

$$x^2 + y^2 = a^6 - 6a^4b^2 + 9a^2b^4 + b^6 - 6a^2b^4 + 9a^4b^2 =$$

$$a^6 + 3a^4b^2 + 3a^2b^4 + b^6 = (a^2 + b^2)^3.$$

Hence

$$(a^2 + b^2)^3 = x^2 + y^2 \leq 47^2 + 52^2.$$

Luckily, we can easily check that $47^2 + 52^2 = 17^3$, giving the desired result.

18. Prove that

$$(1^4 + 1^2 + 1)(2^4 + 2^2 + 1)\ldots(n^4 + n^2 + 1)$$

is not a perfect square if n is a positive integer.

Solution. The key point is the factorization

$$n^4 + n^2 + 1 = n^4 + 2n^2 + 1 - n^2 = (n^2 + 1)^2 - n^2 = (n^2 - n + 1)(n^2 + n + 1).$$

Let

$$x_n = n^2 - n + 1 = (n-1)n + 1.$$

Then
$$n^2 + n + 1 = n(n+1) + 1 = x_{n+1},$$
hence
$$(1^4+1^2+1)(2^4+2^2+1)...(n^4+n^2+1) = (x_1x_2)(x_2x_3)...(x_{n-1}x_n)(x_nx_{n+1}) =$$
$$x_1(x_2...x_n)^2 x_{n+1}.$$

Since $x_1 = 1$, proving the result comes down to proving that $x_{n+1} = n^2 + n + 1$ is not a square. But this is clear, since
$$n^2 < n^2 + n + 1 < (n+1)^2.$$

19. Let x, y be real numbers such that
$$(x + \sqrt{x^2 + 1})(y + \sqrt{y^2 + 1}) = 1.$$
Prove that $x + y = 0$.

Solution. We will exploit multiplication by conjugates. The key point is that
$$(x + \sqrt{x^2 + 1})(x - \sqrt{x^2 + 1}) = -1$$
for all x. Multiplying by the conjugate for x gives
$$y + \sqrt{y^2 + 1} = -x + \sqrt{x^2 + 1}$$
and multiplying by the conjugate for y gives
$$x + \sqrt{x^2 + 1} = -y + \sqrt{y^2 + 1}.$$
Adding these two equations and canceling gives $2(x+y) = 0$ or $x+y = 0$.

20. Prove that for all $x, y, z \geq 0$
$$x^2 + xy^2 + xyz^2 \geq 4xyz - 4.$$

Solution. Let us write the inequality as
$$x^2 + 4 + xy^2 + xyz^2 \geq 4xyz.$$

There is a very easy way to get rid of x: just use that $x^2 + 4 \geq 4x$ (equivalent to $(x-2)^2 \geq 0$). Thus it suffices to prove that
$$4 + y^2 + yz^2 \geq 4yz.$$

Now we repeat the previous step: we write $4 + y^2 \geq 4y$ and reduce the problem to $4 + z^2 \geq 4z$, which itself is equivalent to $(z-2)^2 \geq 0$, thus true.

There is also a rather tricky proof using the AM-GM inequality:

$$4 + x^2 + xy^2 + xyz^2 = 4 + x^2 + \frac{xy^2}{2} + \frac{xy^2}{2} + \frac{xyz^2}{4} + ... + \frac{xyz^2}{4}$$

$$\geq 8\sqrt[8]{4 \cdot x^2 \cdot \frac{x^2 y^4}{4} \cdot \frac{x^4 y^4 z^8}{4^4}} = 4xyz.$$

Finally, we can also try to complete the squares and obtain the equivalent inequality
$$xy(z-2)^2 + x(y-2)^2 + (x-2)^2 \geq 0,$$
which is clear.

21. Let x, y, a be real numbers such that
$$x + y = x^3 + y^3 = x^5 + y^5 = a.$$

Find all possible values of a.

Solution. By hypothesis we have
$$(x+y)(x^5 + y^5) = (x^3 + y^3)^2,$$
which becomes
$$x^6 + xy^5 + x^5 y + y^6 = x^6 + y^6 + 2x^3 y^3,$$
that is
$$xy(x^4 + y^4 - 2x^2 y^2) = 0 \quad \text{or} \quad xy(x^2 - y^2)^2 = 0.$$

Let us discuss a few cases. If $x = 0$, then the equations become $y = y^3 = y^5 = a$. We deduce that $y \in \{-1, 0, 1\}$ and accordingly $a \in \{-1, 0, 1\}$, all of which are possible values of a. The case $y = 0$ is similar and yields the same values of a. So assume that $xy \neq 0$. Then by the previous discussion we must have $x^2 = y^2$, that is either $x = y$ or $x = -y$. If $x = -y$, then all equations reduce to $a = 0$, so suppose that $x = y$. Then the equations become $2x = 2x^3 = 2x^5 = a$. Again, we obtain $x \in \{-1, 0, 1\}$ and then $a \in \{-2, 0, 2\}$, all of which are admissible values. Hence the answer of the problem is $a \in \{-2, -1, 0, 1, 2\}$.

22. Let a and n be positive integers with $n > 1$. Find the integer part of the number $(\sqrt[n]{a} - \sqrt[n]{a+1} + \sqrt[n]{a+2})^n$.

Solution. Since $a + 2 > a + 1$, it is clear that

$$\sqrt[n]{a} - \sqrt[n]{a+1} + \sqrt[n]{a+2} > \sqrt[n]{a},$$

thus

$$(\sqrt[n]{a} - \sqrt[n]{a+1} + \sqrt[n]{a+2})^n > a.$$

We will prove that

$$(\sqrt[n]{a} - \sqrt[n]{a+1} + \sqrt[n]{a+2})^n < a+1, \quad \text{or} \quad \sqrt[n]{a} - \sqrt[n]{a+1} + \sqrt[n]{a+2} < \sqrt[n]{a+1},$$

showing therefore that the answer of the problem is a. To prove the previous inequality, we write it as

$$\sqrt[n]{a+2} - \sqrt[n]{a+1} < \sqrt[n]{a+1} - \sqrt[n]{a}$$

or equivalently

$$\frac{1}{\sqrt[n]{(a+2)^{n-1}} + \ldots + \sqrt[n]{(a+1)^{n-1}}} < \frac{1}{\sqrt[n]{(a+1)^{n-1}} + \ldots + \sqrt[n]{a^{n-1}}}.$$

But this is clear, since

$$\sqrt[n]{(a+2)^{n-1}} > \sqrt[n]{(a+1)^{n-1}}, \quad \ldots, \quad \sqrt[n]{(a+1)^{n-1}} > \sqrt[n]{a^{n-1}}.$$

23. Let $f(x) = (x^2 + x)(x^2 + x + 1) + \frac{x^5}{5}$. Evaluate $f(\sqrt[5]{5} - 1)$.

Solution. We are certainly not supposed to simply insert $x = \sqrt[5]{5} - 1$ in the expression giving $f(x)$, since that would end up in a quite terrible number. Instead, we start by rearranging f a little bit, by expanding and rearranging terms

$$f(x) = x^4 + x^3 + x^2 + x^3 + x^2 + x + \frac{x^5}{5} = x^4 + 2x^3 + 2x^2 + x + \frac{x^5}{5}$$

$$= \frac{x^5 + 5x^4 + 10x^3 + 10x^2 + 5x}{5}.$$

We recognize the successive terms in the expansion of $(x+1)^5$, so we conclude that

$$f(x) = \frac{(x+1)^5 - 1}{5}, \quad \text{thus} \quad f(\sqrt[5]{5} - 1) = \frac{4}{5}.$$

24. Let a, b, c be positive real numbers such that
$$\frac{1}{a} + \frac{1}{b} + \frac{1}{c} = 1.$$
Prove that
$$\frac{1}{a+bc} + \frac{1}{b+ac} + \frac{1}{c+ab} = \frac{2}{(1+\frac{a}{b})(1+\frac{b}{c})(1+\frac{c}{a})}.$$

Solution. Let us start by choosing better variables: let
$$x = \frac{1}{a}, \quad y = \frac{1}{b}, \quad z = \frac{1}{c}.$$
The hypothesis becomes
$$x + y + z = 1.$$
On the other hand,
$$\frac{1}{a+bc} = \frac{1}{\frac{1}{x} + \frac{1}{yz}} = \frac{xyz}{x+yz}$$
and
$$\frac{2}{(1+\frac{a}{b})(1+\frac{b}{c})(1+\frac{c}{a})} = \frac{2xyz}{(x+y)(y+z)(z+x)}.$$
Thus we need to prove that
$$\frac{1}{x+yz} + \frac{1}{y+zx} + \frac{1}{z+xy} = \frac{2}{(x+y)(y+z)(z+x)}.$$
Let us observe that
$$x + yz = x(x+y+z) + yz = x^2 + xy + xz + yz$$
$$= x(x+y) + z(x+y) = (x+z)(x+y).$$
Writing down similar identities, we end up with
$$\frac{1}{x+yz} + \frac{1}{y+zx} + \frac{1}{z+xy}$$
$$= \frac{1}{(x+y)(x+z)} + \frac{1}{(y+x)(y+z)} + \frac{1}{(z+x)(z+y)}$$
$$= \frac{(x+y) + (y+z) + (z+x)}{(x+y)(y+z)(z+x)}$$
$$= \frac{2}{(x+y)(y+z)(z+x)},$$
where the last equality follows from the hypothesis $x + y + z = 1$.

25. Solve the equation
$$\frac{8}{\{x\}} = \frac{9}{x} + \frac{10}{\lfloor x \rfloor}.$$

Solution. Let us denote $a = \{x\}$ and $b = \lfloor x \rfloor$. Then $x = a + b$, so the equation becomes
$$\frac{8}{a} = \frac{9}{a+b} + \frac{10}{b}.$$
Clearing denominators yields the equivalent equation
$$8b(a+b) = 9ab + 10a(a+b),$$
which can be further simplified to
$$10a^2 + 11ab - 8b^2 = 0.$$

This quadratic equation in a and b is homogeneous, so by denoting $c = \frac{a}{b}$ we obtain the quadratic equation $10c^2 + 11c - 8 = 0$, which solved with the usual procedure gives $c = -\frac{8}{5}$ or $c = \frac{1}{2}$. Coming back to our notations, we obtain
$$\{x\} = \frac{1}{2}\lfloor x \rfloor \quad \text{or} \quad \{x\} = -\frac{8}{5}\lfloor x \rfloor.$$

Next, remember that in the statement of the problem we divided by $\lfloor x \rfloor x$, so $\lfloor x \rfloor$ is a nonzero integer. It follows that $-\frac{8}{5}\lfloor x \rfloor$ is certainly not in $[0,1)$. Since $\{x\} \in [0,1)$ for all x, it follows that the equation $\{x\} = -\frac{8}{5}\lfloor x \rfloor$ has no solution. So it remains to solve $\{x\} = \frac{1}{2}\lfloor x \rfloor$. Again, since $\{x\} \in [0,1)$ and $\lfloor x \rfloor$ is a nonzero integer, we must have $\lfloor x \rfloor = 1$ and then $\{x\} = \frac{1}{2}$. Thus $x = 1 + \frac{1}{2} = \frac{3}{2}$, which is the unique solution of the problem.

26. Let a, b, c be positive real numbers not exceeding 1. Prove that
$$(a + b + c)(abc + 1) \geq ab + bc + ca + 3abc.$$

Solution. Write the inequality as
$$abc(a + b + c) + a + b + c \geq ab + bc + ca + 3abc,$$
or equivalently as
$$abc(3 - a - b - c) \leq b(1 - a) + c(1 - b) + a(1 - c).$$

This follows from the fact that $a, b, c \in [0, 1]$, hence
$$abc(1 - a) \leq b(1 - a),$$
since $1 - a \geq 0$ and $0 \leq abc \leq b$. We obtain similarly
$$abc(1 - b) \leq c(1 - b), \quad abc(1 - c) \leq a(1 - c)$$
and the result follows by adding these inequalities.

27. Solve in real numbers the system of equations
$$\begin{cases} (x-2y)(x-4z) = 3 \\ (y-2z)(y-4x) = 5 \\ (z-2x)(z-4y) = -8. \end{cases}$$

Solution. Since there is no apparent symmetry in this system, we start by brutally multiplying out the factors in the left hand-side and adding the resulting equations (one hint being furnished by the fact that $3+5-8=0$ and that all terms in the left hand-side are homogeneous). We end up with
$$x^2 + y^2 + z^2 + 2zx + 2zy + 2yx = 0,$$
which is equivalent to $(x+y+z)^2 = 0$ and then to $x = -y-z$. Replacing this value of x in the first and the second equations, we obtain the system in the unknowns y, z
$$\begin{cases} 3y^2 + 16yz + 5z^2 = 3 \\ 5y^2 - 6zy - 8z^2 = 5. \end{cases}$$
Next, using that $3 \cdot 5 - 5 \cdot 3 = 0$, we multiply the first equation by 5, the second one by 3 and we subtract the resulting relations. This gives
$$29z^2 + 58zy = 0, \quad \text{that is} \quad z(z+2y) = 0.$$
Let us discuss two cases. If $z = 0$, then the previous system becomes $3y^2 = 3$ and $5y^2 = 5$, having two solutions $y = \pm 1$. Since $x = -y - z$, we obtain two solutions $(x, y, z) = \{(1, -1, 0), (-1, 1, 0)\}$ of the initial system. If on the other hand $z = -2y$, then the first equation of the previous system becomes
$$3y^2 - 32y^2 + 20y^2 = 3,$$
that is $-9y^2 = 3$, and this clearly has no real solution. We conclude that this second case is impossible, hence the system has only the solutions found above.

28. If a, b, c, d are real numbers such that $a^2 + b^2 + (a+b)^2 = c^2 + d^2 + (c+d)^2$, prove that
$$a^4 + b^4 + (a+b)^4 = c^4 + d^4 + (c+d)^4.$$

Solution. Let us start by writing the hypothesis in a simpler form: we expand, rearrange terms and divide by 2, obtaining equivalent forms of the hypothesis
$$a^2 + b^2 + a^2 + 2ab + b^2 = c^2 + d^2 + c^2 + 2cd + d^2 \quad \text{or} \quad a^2 + ab + b^2 = c^2 + cd + d^2.$$

Let us do the same thing with the conclusion. Here it will be useful to recall the binomial formula

$$(a+b)^4 = a^4 + 4a^3b + 6a^2b^2 + 4ab^3 + b^4.$$

Hence the conclusion can be written as

$$2(a^4 + b^4) + 4a^3b + 6a^2b^2 + 4ab^3 = 2(c^4 + d^4) + 4c^3d + 4cd^3 + 6c^2d^2,$$

or equivalently, after division by 2, as

$$a^4 + 2a^3b + 3a^2b^2 + 2ab^3 + b^4 = c^4 + 2c^3d + 3c^2d^2 + 2cd^3 + d^4.$$

Well, both terms in the previous equality look very close to the square of $(a^2 + ab + b^2)^2$ and $(c^2 + cd + d^2)^2$. And not only do the look close, they are actually equal to these numbers! So the conclusion follows simply by squaring the equality $a^2 + ab + b^2 = c^2 + cd + d^2$.

Note that what the proof really shows is that for all real numbers a, b we have
$$\frac{a^4 + b^4 + (a+b)^4}{2} = \left(\frac{a^2 + b^2 + (a+b)^2}{2}\right)^2.$$

29. The polynomial $P(x) = x^3 + ax^2 + bx + c$ has three distinct real roots. The polynomial $P(Q(x))$, where $Q(x) = x^2 + x + 2001$, has no real roots. Prove that $P(2001) > \frac{1}{64}$.

Solution. Let x_1, x_2, x_3 be the roots of the polynomial P, so that

$$P(x) = (x - x_1)(x - x_2)(x - x_3)$$

for all x. Thus

$$P(Q(x)) = (x^2 + x + 2001 - x_1)(x^2 + x + 2001 - x_2)(x^2 + x + 2001 - x_3).$$

By hypothesis, this last expression is always nonzero. So the equation $x^2 + x + 2001 - x_i = 0$ has no real solution for $i = 1, 2, 3$. We deduce that the corresponding discriminant is negative, so that

$$2001 - x_i > \frac{1}{4}, \quad \text{for} \quad i = 1, 2, 3.$$

Finally, we obtain

$$P(2001) = (2001 - x_1)(2001 - x_2)(2001 - x_3) > \frac{1}{4^3} = \frac{1}{64}.$$

30. A sequence $(a_n)_{n \geq 0}$ is defined by $a_0 = 1$ and

$$a_{n+1} = a_{\lfloor \frac{7n}{9} \rfloor} + a_{\lfloor \frac{n}{9} \rfloor}$$

for $n \geq 0$. Prove that there is an n such that $a_n < \frac{n}{2001!}$.

Solution. Note that $a_1 = 2$. We prove by strong induction that $a_n \leq 2n$ for $n \geq 1$. If $n = 1$ this has already been established. Next, we easily compute using the recurrence relation

$$a_2 = 2, \quad a_3 = a_4 = 3, \quad a_5 = a_6 = a_7 = 4, \quad a_8 = 5, a_9 = 5$$

so $a_n \leq 2n$ for $1 \leq n \leq 9$. Assume that the inequality $a_k \leq 2k$ holds up to $n \geq 9$ (included). Then

$$a_{n+1} = a_{\lfloor \frac{7n}{9} \rfloor} + a_{\lfloor \frac{n}{9} \rfloor} \leq 2\lfloor \frac{7n}{9} \rfloor + 2\lfloor \frac{n}{9} \rfloor \leq 2 \cdot \frac{8n}{9} \leq 2(n+1)$$

and the inductive step is proved.

Next, if $n \geq 9$, then $\lfloor \frac{n}{9} \rfloor$ and $\lfloor \frac{7n}{9} \rfloor \geq 1$, hence with the same arguments as above we obtain

$$a_{n+1} = a_{\lfloor \frac{7n}{9} \rfloor} + a_{\lfloor \frac{n}{9} \rfloor} \leq 2 \cdot \frac{8n}{9}.$$

We conclude that $a_n \leq 2 \cdot \frac{8}{9} n$ for $n \geq 10$. Thus for $n \geq 90$ we have

$$a_{n+1} \leq 2 \cdot \frac{8}{9} \cdot \left(\lfloor \frac{7n}{9} \rfloor + \lfloor \frac{n}{9} \rfloor \right) \leq 2 \frac{8^2}{9^2} n$$

and so $a_n \leq 2 \cdot \frac{8^2}{9^2} n$ for $n \geq 91$. An immediate induction then shows that $a_n \leq 2 \cdot \frac{8^k}{9^k} n$ for $n \geq x_k$, where $x_0 = 1$ and $x_{k+1} = 9x_k + 1$. This being said, since $\frac{8}{9} < 1$, we can find k such that $\frac{8^k}{9^k} < \frac{1}{2 \cdot 2001!}$. Then the previous discussion shows that $a_n < \frac{n}{2001!}$ for all $n \geq x_k$ and we are done.

31. Prove that if real numbers a, b satisfy

$$a^2(2a^2 - 4ab + 3b^2) = 3 \quad \text{and} \quad b^2(3a^2 - 4ab + 2b^2) = 5,$$

then $a^3 + b^3 = 2(a + b)$.

Solution. It is much easier to get some information by analyzing the conclusion first, rather than the quite complicated hypothesis. The reason is that $a^3 + b^3 = 2(a+b)$ is equivalent to $(a+b)(a^2 - ab + b^2) = 2(a+b)$, so to $a + b = 0$ or $a^2 - ab + b^2 = 2$. If $a + b = 0$, then replacing $b = -a$ in the hypothesis yields a contradiction. So what we really have to prove is that $a^2 - ab + b^2 = 2$. Now, the hypothesis is written in terms of fourth degree polynomials in a and b, so it is natural to express the conclusion in the same way. This means of course squaring $a^2 - ab + b^2$. Hence the conclusion is equivalent (because

$$a^2 - ab + b^2 = \left(a - \frac{b}{2}\right)^2 + \frac{3b^2}{4}$$

is nonnegative) to

$$4 = (a^2 - ab + b^2)^2 = a^4 + a^2b^2 + b^4 - 2a^3b - 2ab^3 + 2a^2b^2$$

or equivalently

$$a^4 - 2a^3b + 3a^2b^2 - 2ab^3 + b^4 = 4.$$

On the other hand, the hypothesis can be written as

$$2a^4 - 4a^3b + 3a^2b^2 = 3 \quad \text{and} \quad 3a^2b^2 - 4ab^3 + 2b^4 = 5.$$

Well, the miracle is that by adding these two equations and dividing by 2 we obtain exactly the previously desired relation.

Here is an alternative approach: we start with adding the two given equations:

$$2a^4 - 4a^3b + 6a^2b^2 - 4ab^4 + 2b^4 = 8.$$

It is tempting to divide by 2, but it is better to avoid doing that and rewrite the left hand-side as $a^4 + b^4 + (a-b)^4$, using the binomial formula. Hence

$$a^4 + b^4 + (a - b)^4 = 8.$$

Now we use the identity

$$\frac{a^4 + b^4 + (a+b)^4}{2} = \left(\frac{a^2 + b^2 + (a+b)^2}{2}\right)^2$$

established in the solution of exercise 28. It follows that

$$a^2 - ab + b^2 = \frac{a^2 + b^2 + (a - b)^2}{2} = 2$$

and multiplying by $a + b$ gives the desired result.

32. If x and y are real numbers such that $\frac{x^2+y^2}{x^2-y^2} + \frac{x^2-y^2}{x^2+y^2} = k$, express

$$\frac{x^8 + y^8}{x^8 - y^8} + \frac{x^8 - y^8}{x^8 + y^8}$$

in terms of k.

Solution. Putting it over a common denominator we see that
$$k = \frac{(x^2+y^2)^2 + (x^2-y^2)^2}{(x^2+y^2)(x^2-y^2)} = 2 \cdot \frac{x^4+y^4}{x^4-y^4}.$$

Noticing that this looks like part of the formula for k, we use this result again to obtain
$$\frac{k}{2} + \frac{2}{k} = \frac{x^4+y^4}{x^4-y^4} + \frac{x^4-y^4}{x^4+y^4} = 2 \cdot \frac{x^8+y^8}{x^8-y^8}.$$

Hence
$$\frac{x^8+y^8}{x^8-y^8} + \frac{x^8-y^8}{x^8+y^8} = \frac{k}{4} + \frac{1}{k} + \frac{2}{\frac{k}{2}+\frac{2}{k}} = \frac{k}{4} + \frac{1}{k} + \frac{4k}{k^2+4}.$$

Here is an alternative approach. The key point is that the hypothesis and the conclusion do not change if we multiply x and y by the same number. Also, what really matters here is x^2 and y^2, and not x, y. This will allow us to work with only one variable $t = \frac{x^2}{y^2}$. Well, to be precise, this would work if y were nonzero. So, let us start with the easy case $y = 0$. Then the hypothesis becomes simply $2 = k$, and
$$\frac{x^8+y^8}{x^8-y^8} + \frac{x^8-y^8}{x^8+y^8} = 2.$$

Suppose now that $y \neq 0$ and set $t = \frac{x^2}{y^2}$. Then $x^2 = ty^2$, so the hypothesis becomes
$$\frac{x^2(t+1)}{x^2(t-1)} + \frac{x^2(t-1)}{x^2(t+1)} = k$$
or equivalently
$$\frac{t+1}{t-1} + \frac{t-1}{t+1} = k.$$

Let us try finding t explicitly in terms of k. We clear denominators in the previous relation and rewrite is as
$$(t+1)^2 + (t-1)^2 = k(t^2-1), \quad \text{or} \quad 2(t^2+1) = k(t^2-1),$$

and finally as $t^2(k-2) = k+2$. Thus $k \neq 2$ (otherwise we would also get $k = -2$) and
$$t^2 = \frac{k+2}{k-2}.$$

Solutions to introductory problems 143

We are asked to compute
$$\frac{x^8+y^8}{x^8-y^8}+\frac{x^8-y^8}{x^8+y^8}=\frac{t^4+1}{t^4-1}+\frac{t^4-1}{t^4+1}.$$

Since $t^2=\frac{k+2}{k-2}$, we have
$$t^4-1=\frac{(k+2)^2}{(k-2)^2}-1=\frac{k^2+4k+4-(k^2-4k+4)}{(k-2)^2}=\frac{8k}{(k-2)^2}$$

and
$$t^4+1=\frac{(k+2)^2}{(k-2)^2}+1=\frac{k^2+4k+4+(k^2-4k+4)}{(k-2)^2}=\frac{2k^2+8}{(k-2)^2}.$$

We conclude that
$$\frac{t^4-1}{t^4+1}+\frac{t^4+1}{t^4-1}=\frac{8k}{8+2k^2}+\frac{8+2k^2}{8k}=\frac{4k}{4+k^2}+\frac{4+k^2}{4k}.$$

Note that if $y=0$, then we get the same answer (since the last quantity equals 2 when evaluated at $k=2$), so we conclude that
$$\frac{x^8+y^8}{x^8-y^8}+\frac{x^8-y^8}{x^8+y^8}=\frac{4k}{4+k^2}+\frac{4+k^2}{4k}$$

in all cases.

33. Find all positive real numbers x, y, z satisfying
$$x+\frac{y}{z}=y+\frac{z}{x}=z+\frac{x}{y}=2.$$

Solution. The assumption that x,y,z are positive should be considered as a hint that inequalities play an important role in this problem. Let us add the equations:
$$x+y+z+\frac{y}{z}+\frac{z}{x}+\frac{x}{y}=6.$$

By the AM-GM inequality we have
$$\frac{y}{z}+\frac{z}{x}+\frac{x}{y}\geq 3,$$

hence we must have $x+y+z\leq 3$. On the other hand, writing the equations as
$$xz+y=2z,\quad xy+z=2x,\quad yz+x=2y$$

and adding them up yields
$$xy + yz + zx = x + y + z.$$

On the other hand, since $x^2 + y^2 + z^2 \geq xy + yz + zx$, we obtain
$$x + y + z = xy + yz + zx \leq \frac{(x+y+z)^2}{3},$$
thus $x + y + z \geq 3$. Since we have already obtained $x + y + z \leq 3$, we deduce that $x + y + z = 3$ and moreover that all previous inequalities must be equalities, which is the case if and only if $x = y = z$. We conclude that $x = y = z = 1$ is the unique solution of the problem.

34. Monic quadratic polynomials $f(x)$ and $g(x)$ are such that the equations $f(g(x)) = 0$ and $g(f(x)) = 0$ have no real solutions. Prove that at least one of equations $f(f(x)) = 0$ and $g(g(x)) = 0$ has no real solution (recall that a polynomial is monic if the leading coefficient equals 1).

Solution. Let us argue by contradiction and assume that both equations $f(f(x)) = 0$ and $g(g(x)) = 0$ have at least one real root. Hence there is a root a of f such that the equation $f(x) = a$ has a real solution. Therefore the discriminant of the polynomial $f - a$ is nonnegative. Let Δ_1 be the discriminant of f. Then this says $\Delta_1 + 4a \geq 0$. Similarly, there is a root b of g such that $g(x) = b$ has a real solution. If we let Δ_2 be the discriminant of g, then we conclude $\Delta_2 + 4b \geq 0$.

However, since $f(g(x))$ and $g(f(x))$ have no real solutions, the equations $g(x) - a$ and $f(x) - b$ must not have real solutions. Hence their discriminants must be negative. That is, $\Delta_2 + 4a < 0$ and $\Delta_1 + 4b < 0$. Adding the first two inequalities gives $\Delta_1 + \Delta_2 + 4a + 4b \geq 0$ and adding the last two gives $\Delta_1 + \Delta_2 + 4a + 4b < 0$, a contradiction.

35. Prove that for all $x, y > 0$
$$\frac{2xy}{x+y} + \sqrt{\frac{x^2+y^2}{2}} \geq \frac{x+y}{2} + \sqrt{xy}.$$

Solution. We will change the inequality a little bit, writing it as
$$\sqrt{\frac{x^2+y^2}{2}} - \sqrt{xy} \geq \frac{x+y}{2} - \frac{2xy}{x+y}.$$

Next, we work out each term:
$$\sqrt{\frac{x^2+y^2}{2}} - \sqrt{xy} = \frac{\frac{x^2+y^2}{2} - xy}{\sqrt{xy} + \sqrt{\frac{x^2+y^2}{2}}} = \frac{(x-y)^2}{2(\sqrt{xy} + \sqrt{\frac{x^2+y^2}{2}})}$$

and
$$\frac{x+y}{2} - \frac{2xy}{x+y} = \frac{(x+y)^2 - 4xy}{2(x+y)} = \frac{(x-y)^2}{2(x+y)}.$$

Thus the inequality is equivalent to

$$\frac{1}{2(\sqrt{xy} + \sqrt{\frac{x^2+y^2}{2}})} \geq \frac{1}{2(x+y)},$$

or

$$\sqrt{xy} + \sqrt{\frac{x^2+y^2}{2}} \leq x+y.$$

This is a quite easy application of the Cauchy-Schwarz inequality:

$$\left(\sqrt{xy} + \sqrt{\frac{x^2+y^2}{2}}\right)^2 \leq 2\left(\frac{x^2+y^2}{2} + xy\right) = (x+y)^2.$$

36. A quadratic polynomial P has the property that $P(x^3 + x) \geq P(x^2 + 1)$ for all real numbers x. Find the sum of the roots of P.

   Solution. Let $P(x) = ax^2 + bx + c$. Then

   $$P(x^3+x) - P(x^2+1) = ax^2(x^2+1)^2 + bx(x^2+1) - a(x^2+1)^2 - b(x^2+1) =$$

   $$a(x^2+1)^2(x^2-1) + b(x-1)(x^2+1) = (x-1)(x^2+1)(a(x+1)(x^2+1)+b).$$

   Since this is always nonnegative, we have

   $$(x-1)(a(x+1)(x^2+1)+b) \geq 0$$

   for all x. Taking $x = 1 + u$, we deduce that $a(u+2)(1+(1+u)^2) + b$ has the same sign as u for all $u \neq 0$. When u is negative and very close to 0, the number $a(u+2)(1+(1+u)^2)+b$ is very close to $4a+b$, so $4a+b$ must be nonpositive. Similarly, taking u positive and very close to 0, we obtain that $4a+b$ must be nonnegative. Thus $4a+b=0$ and so the sum of roots of P is $-\frac{b}{a} = 4$.

   Remark 14.1. The crucial fact used by this problem is that if $Q(x)$ is a polynomial with $Q(x) \geq 0$ for all real x, then every root of Q has even multiplicity. To see this note that if $x = a$ is a root with odd multiplicity m, then we can write $Q(x) = (x-a)^m G(x)$ for some polynomial G with $G(a) \neq 0$. Hence $G(x)$ will have the same sign as $G(a)$ on some interval centered at a. But $(x-a)^m$ will be positive for $x > a$ and negative for $x < a$, hence Q will be negative on one of the sides of a, a contradiction.

In the current problem, we let $P(x) = ax^2 + bx + c$ and we take

$$Q(x) = P(x^3 + x) - P(x^2 + 1) =$$

$$a(x^3 + x)^2 + b(x^3 + x) - a(x^2 + 1)^2 - b(x^2 + 1) =$$

$$ax^6 + ax^4 + bx^3 - (a+b)x^2 + bx - a - b.$$

From the definition, $Q(1) = P(2) - P(2) = 0$, so $x = 1$ is a root and we factor to obtain

$$Q(x) = (x-1)(ax^5 + ax^4 + 2ax^3 + (2a+b)x^2 + ax + a + b).$$

By the result above $Q(x) \geq 0$ implies that $x = 1$ must be a double root of Q and hence $8a + 2b = 0$ and the sum of the roots is $-\frac{b}{a} = 4$.

37. Solve the equation

$$x^2 - 10x + 1 = \sqrt{x}(x+1).$$

Solution. Let x be a solution of the equation. Then $x \geq 0$ and actually $x > 0$, since $x = 0$ is clearly not a solution. Now, we're taking an indirect approach: we divide by x to obtain the equivalent equation

$$x - 10 + \frac{1}{x} = \sqrt{x} + \frac{1}{\sqrt{x}}.$$

Now, we make the substitution $y = \sqrt{x} + \frac{1}{\sqrt{x}}$, the key point being that $x + \frac{1}{x} + 2 = y^2$. Thus the equation can also be written as $y^2 - 12 = y$. This quadratic equation is easily solved and we obtain the two solutions $y = 4$ and $y = -3$. However, $y > 0$ by definition, thus $y = -3$ is not acceptable. Hence $y = 4$ and we have $x + \frac{1}{x} = y^2 - 2 = 14$. Solving the new quadratic equation $x^2 - 14x + 1 = 0$ finally gives the solutions $7 \pm 4\sqrt{3}$, both of which are solutions of the original equation, since both are positive.

38. The polynomial $x^3 + ax^2 + bx + c$ has three real roots. Prove that if $-2 \leq a + b + c \leq 0$, then at least one of these roots belongs to $[0, 2]$.

Solution. Let x_1, x_2, x_3 be the roots of the polynomial, so that

$$x^3 + ax^2 + bx + c = (x - x_1)(x - x_2)(x - x_3)$$

for all x. Taking $x = 1$, we obtain

$$1 + a + b + c = (1 - x_1)(1 - x_2)(1 - x_3).$$

Combined with the hypothesis, this yields

$$|1-x_1| \cdot |1-x_2| \cdot |1-x_3| = |1+a+b+c| \le 1.$$

Thus the smallest of the numbers $|1-x_1|, |1-x_2|, |1-x_3|$, say $|1-x_1|$ without loss of generality, does not exceed 1. It follows that $x_1 \in [0,2]$ and we are done.

39. Consider the system of equations

$$\begin{cases} a_{11}x_1 + a_{12}x_2 + a_{13}x_3 = 0 \\ a_{21}x_1 + a_{22}x_2 + a_{23}x_3 = 0. \\ a_{31}x_1 + a_{32}x_2 + a_{33}x_3 = 0 \end{cases}$$

with unknowns x_1, x_2, x_3. Suppose that:

a) a_{11}, a_{22}, a_{33} are positive numbers;

b) the remaining coefficients are negative numbers;

c) in each equation, the sum of the coefficients is positive.

Prove that the given system has only the solution $x_1 = x_2 = x_3 = 0$.

Solution. Suppose that (x_1, x_2, x_3) is a solution of the system. By symmetry we may assume that $|x_1|$ is maximal among $|x_1|, |x_2|$ and $|x_3|$. From the first equation we obtain

$$|a_{11}x_1| = |a_{12}x_2 + a_{13}x_3| \le |a_{12}| \cdot |x_2| + |a_{13}| \cdot |x_3| \le (|a_{12}| + |a_{13}|)|x_1|,$$

hence

$$(|a_{11}| - |a_{12}| - |a_{13}|)|x_1| \le 0.$$

Now, the hypothesis ensures that

$$|a_{11}| - |a_{12}| - |a_{13}| = a_{11} + a_{12} + a_{13} > 0,$$

so the previous inequality yields $|x_1| \le 0$ and then $x_1 = 0$. Since $|x_1|$ is maximal among $|x_1|, |x_2|$ and $|x_3|$, it follows that $x_1 = x_2 = x_3 = 0$ and we are done.

40. Find all triples (x, y, z) of positive real numbers for which there is a positive real number t such that the following inequalities hold simultaneously:

$$\frac{1}{x} + \frac{1}{y} + \frac{1}{z} + t \le 4, \quad x^2 + y^2 + z^2 + \frac{2}{t} \le 5.$$

Solution. Let x, y, z, t satisfying the inequalities in the statement. Then adding the inequalities

$$8 \geq 2t + \frac{2}{x} + \frac{2}{y} + \frac{2}{z}$$

and $5 \geq \frac{2}{t} + x^2 + y^2 + z^2$ yields

$$13 \geq 2\left(t + \frac{1}{t}\right) + x^2 + \frac{2}{x} + y^2 + \frac{2}{y} + z^2 + \frac{2}{z}.$$

Next, the AM-GM inequality yields

$$x^2 + \frac{2}{x} = x^2 + \frac{1}{x} + \frac{1}{x} \geq 3\sqrt[3]{x^2 \cdot \frac{1}{x^2}} = 3.$$

Writing down similar inequalities and adding them gives

$$13 \geq 9 + 2\left(t + \frac{1}{t}\right).$$

We conclude that $t + \frac{1}{t} \leq 2$, thus $(t-1)^2 \leq 0$. Not only does this imply $t = 1$, but it also implies that all previous inequalities were actually equalities. In particular, we must have equality when applying AM-GM, which happens precisely when $x = y = z = 1$. We conclude that $(x, y, z) = (1, 1, 1)$ is the unique solution of the problem (in this case we can take $t = 1$).

41. Prove that if $a + b + c = 0$, then

$$\left(\frac{a}{b-c} + \frac{b}{c-a} + \frac{c}{a-b}\right)\left(\frac{b-c}{a} + \frac{c-a}{b} + \frac{a-b}{c}\right) = 9.$$

Solution. Let us deal with each factor separately. We have

$$\frac{a}{b-c} + \frac{b}{c-a} + \frac{c}{a-b}$$
$$= \frac{a(c-a)(a-b) + b(b-c)(a-b) + c(b-c)(c-a)}{(a-b)(b-c)(c-a)}$$

and expanding the numerator and rearranging terms yields

$$a(c-a)(a-b) + b(b-c)(a-b) + c(b-c)(c-a)$$
$$= a^2(b+c) + b^2(c+a) + c^2(a+b) - (a^3 + b^3 + c^3 + 3abc).$$

Since $a+b+c=0$, we have
$$a^3+b^3+c^3 = 3abc$$
and
$$a^2(b+c)+b^2(c+a)+c^2(a+b) = -a^3-b^3-c^3 = -3abc.$$
Hence we conclude that
$$\frac{a}{b-c}+\frac{b}{c-a}+\frac{c}{a-b} = -\frac{9abc}{(a-b)(b-c)(c-a)}.$$
The second factor is much easier to deal with, since replacing $c-a = -((a-b)+(b-c))$ yields
$$\frac{b-c}{a}+\frac{c-a}{b}+\frac{a-b}{c} = \frac{b-c}{a}-\frac{a-b}{b}-\frac{b-c}{b}+\frac{a-b}{c} =$$
$$(b-c)\left(\frac{1}{a}-\frac{1}{b}\right)+(a-b)\left(\frac{1}{c}-\frac{1}{b}\right) =$$
$$-\frac{(a-b)(b-c)}{ab}+\frac{(a-b)(b-c)}{bc} = -\frac{(a-b)(b-c)(c-a)}{abc}.$$
Combining the two previous relations yields the desired result.

42. Let a,b,c,x,y,z,m,n be positive real numbers satisfying
$$\sqrt[3]{a}+\sqrt[3]{b}+\sqrt[3]{c} = \sqrt[3]{m}, \quad \sqrt{x}+\sqrt{y}+\sqrt{z} = \sqrt{n}.$$
Prove that
$$\frac{a}{x}+\frac{b}{y}+\frac{c}{z} \geq \frac{m}{n}.$$

Solution. Let us denote
$$A = \sqrt[3]{a}, \quad B = \sqrt[3]{b}, \quad C = \sqrt[3]{c}$$
and
$$X = \sqrt{x}, \quad Y = \sqrt{y}, \quad Z = \sqrt{c}.$$
Then the hypothesis can be written
$$m = (A+B+C)^3, \quad n = (X+Y+Z)^2$$
and so the conclusion is equivalent to
$$\frac{A^3}{X^2}+\frac{B^3}{Y^2}+\frac{C^3}{Z^2} \geq \frac{(A+B+C)^3}{(X+Y+Z)^2}.$$

This is Hölder's inequality in disguise, since

$$(X+Y+Z)(X+Y+Z)\left(\frac{A^3}{X^2}+\frac{B^3}{Y^2}+\frac{C^3}{Z^2}\right)$$

$$\geq \left(\sqrt[3]{X^2 \cdot \frac{A^3}{X^2}}+\sqrt[3]{Y^2 \cdot \frac{B^3}{Y^2}}+\sqrt[3]{Z^2 \cdot \frac{C^3}{Z^2}}\right)^3$$

$$= (A+B+C)^3.$$

43. Find all triples x, y, z of positive integers such that

$$x^2y + y^2z + z^2x = xy^2 + yz^2 + zx^2 = 111.$$

Solution. The key relation is

$$x^2y + y^2z + z^2x = xy^2 + yz^2 + zx^2,$$

which can be written as

$$xy(x-y) + yz(y-z) + zx(z-x) = 0.$$

Since $z - x = -((x-y) + (y-z))$, we can rewrite the previous relation as

$$xy(x-y) + yz(y-z) - zx(x-y) - zx(y-z) = 0,$$

that is

$$(xy - zx)(x-y) + (yz - zx)(y-z) = 0, \quad \text{or} \quad (x-y)(y-z)(z-x) = 0.$$

Thus the first equation is equivalent to $(x-y)(y-z)(z-x) = 0$, that is to say two of the numbers x, y, z are equal. Without loss of generality $y = z$. Then the second equation becomes

$$xy^2 + y^3 + yx^2 = 111, \quad \text{or} \quad y(xy + y^2 + x^2) = 111.$$

We factor $111 = 3 \cdot 37$ and we observe that $xy + y^2 + x^2 > y$, since x, y, z are positive integers. Thus we must have $y = 1$ or $y = 3$. If $y = 1$, then $x^2 + x + 1 = 111$, which can also be written as $(x-10)(x+11) = 0$, with the solution $x = 10$ in positive integers. If $y = 3$, then $x^2 + 3x + 9 = 37$, which can be written as $(x-4)(x+7) = 0$ and has the solution $x = 4$. All in all, the solutions are $(x, y, z) = \{(10, 1, 1), (4, 3, 3)\}$ and their permutations.

44. Let a, b, c be nonzero real numbers such that

$$\frac{1}{a^3} + \frac{1}{b^3} + \frac{1}{c^3} = \frac{1}{a^3 + b^3 + c^3}.$$

Prove that
$$\frac{1}{a^5}+\frac{1}{b^5}+\frac{1}{c^5}=\frac{1}{a^5+b^5+c^5}.$$

Solution. Contemplating the two relations in the statement of the problem, a natural question appears: when do we have
$$\frac{1}{x}+\frac{1}{y}+\frac{1}{z}=\frac{1}{x+y+z},$$
for some nonzero real numbers x, y, z? This happens if and only if
$$\frac{1}{x}+\frac{1}{y}=\frac{1}{x+y+z}-\frac{1}{z},$$
which can be also written as
$$\frac{x+y}{xy}=-\frac{x+y}{z(x+y+z)}.$$

This already shows that if $x+y=0$, then the relation holds. Suppose that $x+y\neq 0$. The previous relation becomes
$$\frac{1}{xy}=-\frac{1}{z(x+y+z)}, \quad \text{or} \quad zx+zy+z^2+xy=0.$$

Rearranging terms in the last relation yields the equivalent one $(z+x)(z+y)=0$. We conclude that $\frac{1}{x}+\frac{1}{y}+\frac{1}{z}=\frac{1}{x+y+z}$ if and only if two of the numbers x, y, z add up to 0.

It is now very easy to solve the problem. The hypothesis tells us that two of the numbers a^3, b^3, c^3 add up to 0, say $a^3+b^3=0$. Since a, b are real numbers, this forces $a=-b$. But then $a^5+b^5=0$, hence
$$\frac{1}{a^5}+\frac{1}{b^5}+\frac{1}{c^5}=\frac{1}{a^5+b^5+c^5}$$
and we are done.

45. Solve in nonzero real numbers the system of equations
$$\begin{cases}\frac{7}{x}-\frac{27}{y}=2x^2\\ \frac{9}{y}-\frac{21}{x}=2y^2.\end{cases}$$

Solution. First, we will consider the system as a system with the unknowns $\frac{1}{x}$ and $\frac{1}{y}$. This is a linear system, so rather easy to solve. We obtain
$$\frac{-28}{x}=x^2+3y^2 \quad \text{and} \quad \frac{-36}{y}=3x^2+y^2.$$

Multiplying the first equation by x and the second one by y we obtain the equivalent system
$$\begin{cases} x^3 + 3xy^2 = -28 \\ 3x^2y + y^3 = -36. \end{cases}$$

We recognize the terms of the expansion of $(x+y)^3$ and $(x-y)^3$. By adding and subtracting the equations, we obtain another form of the system
$$\begin{cases} (x+y)^3 = -64 \\ (x-y)^3 = 8. \end{cases}$$

This is quite easy to solve, since it is equivalent to $x+y = -4$ and $x - y = 2$, then $x = -1$ and $y = -3$.

46. Compute the integer part of the number
$$S = \sqrt{2} + \sqrt[3]{\frac{3}{2}} + \ldots + \sqrt[2013]{\frac{2013}{2012}}.$$

Solution. We have
$$S = \sum_{k=1}^{2012} \sqrt[k+1]{1 + \frac{1}{k}}.$$

On the other hand, we have
$$1 < \sqrt[k+1]{1 + \frac{1}{k}} \le 1 + \frac{1}{k(k+1)},$$

the last inequality being a consequence of Bernoulli's inequality (which is simply the binomial formula in this specific case)
$$\left(1 + \frac{1}{k(k+1)}\right)^{k+1} \ge 1 + (k+1) \cdot \frac{1}{k(k+1)} = 1 + \frac{1}{k}.$$

Hence
$$2012 < S < 2012 + \sum_{k=1}^{2012} \frac{1}{k(k+1)}.$$

Since
$$\sum_{k=1}^{2012} \frac{1}{k(k+1)} = \sum_{k=1}^{2012} \left(\frac{1}{k} - \frac{1}{k+1}\right) = 1 - \frac{1}{2013} < 1,$$

we conclude that $\lfloor S \rfloor = 2012$.

47. Prove that for all $a, b, c > 0$

$$\frac{a^3}{b^2+c^2} + \frac{b^3}{c^2+a^2} + \frac{c^3}{a^2+b^2} \geq \frac{a+b+c}{2}.$$

Solution. Writing

$$\frac{a^3}{b^2+c^2} - \frac{a}{2} = \frac{a(2a^2-b^2-c^2)}{2(b^2+c^2)} = \frac{a(a^2-b^2)}{2(b^2+c^2)} + \frac{a(a^2-c^2)}{2(b^2+c^2)},$$

and regrouping terms according to their numerators, we see that

$$\frac{a^3}{b^2+c^2} + \frac{b^3}{c^2+a^2} + \frac{c^3}{a^2+b^2} - \frac{a+b+c}{2}$$
$$= \sum_{\text{cyc}} \left(\frac{a(a^2-b^2)}{2(b^2+c^2)} + \frac{b(b^2-a^2)}{2(a^2+c^2)} \right)$$
$$= \sum_{\text{cyc}} (a^2-b^2) \left(\frac{a}{2(b^2+c^2)} - \frac{b}{2(a^2+c^2)} \right).$$

The last expression clearly vanishes if $a = b$, hence we can put the terms over a common denominator and factor out $a - b$. This gives

$$\frac{a}{2(b^2+c^2)} - \frac{b}{2(a^2+c^2)} = \frac{a(a^2+c^2) - b(b^2+c^2)}{2(a^2+c^2)(b^2+c^2)}$$
$$= \frac{(a-b)(a^2+ab+b^2+c^2)}{2(a^2+c^2)(b^2+c^2)}.$$

Plugging this into the formula above and noting that $(a+b)(a^2+ab+b^2+c^2) > 0$ gives

$$\frac{a^3}{b^2+c^2} + \frac{b^3}{c^2+a^2} + \frac{c^3}{a^2+b^2} - \frac{a+b+c}{2}$$
$$= \sum_{\text{cyc}} \frac{(a-b)^2(a+b)(a^2+ab+b^2+c^2)}{2(a^2+c^2)(b^2+c^2)} \geq 0,$$

with equality if and only if $a = b = c$.

48. Solve in real numbers the equation $x^3 + 1 = 2\sqrt[3]{2x-1}$.

Solution. This is a little bit difficult to solve directly, so let us introduce a new variable $y = \sqrt[3]{2x-1}$. Then $x^3 + 1 = 2y$ and $y^3 = 2x - 1$, that is $y^3 + 1 = 2x$. Suppose that $x > y$. Then $x^3 + 1 > y^3 + 1$, that is $2y > 2x$ and then $y > x$, a contradiction. Similarly, we prove that we cannot have $x < y$, hence $x = y$ and then $x^3 - 2x + 1 = 0$. There is an obvious solution $x = 1$, so we can factor out $x - 1$:

$$x^3 - 2x + 1 = x^3 - x - (x - 1) = (x - 1)(x^2 + x - 1).$$

The solutions of the equation $x^2 + x - 1 = 0$ are given by $\frac{-1 \pm \sqrt{5}}{2}$, hence we conclude that the initial equation has three solutions, namely $1, \frac{-1 \pm \sqrt{5}}{2}$.

49. Prove that for all natural numbers n,

$$\sum_{k=1}^{n^2} \left\{\sqrt{k}\right\} \le \frac{n^2 - 1}{2}.$$

Solution. We will prove this by induction on n, the case $n = 1$ being clear. Suppose that the result holds for n and let us prove it for $n + 1$. Since

$$\sum_{k=1}^{(n+1)^2} \left\{\sqrt{k}\right\} = \sum_{k=1}^{n^2} \left\{\sqrt{k}\right\} + \sum_{k=n^2+1}^{(n+1)^2} \left\{\sqrt{k}\right\},$$

it suffices to prove that

$$\sum_{k=n^2+1}^{(n+1)^2} \left\{\sqrt{k}\right\} \le \frac{(n+1)^2 - 1 - (n^2 - 1)}{2} = \frac{2n+1}{2}.$$

Now, if $n^2 + 1 \le k < (n+1)^2$, then $\lfloor \sqrt{k} \rfloor = n$, while for $k = (n+1)^2$ we have $\left\{\sqrt{k}\right\} = 0$. Thus it suffices to prove that

$$\sum_{k=n^2+1}^{(n+1)^2-1} \left(\sqrt{k} - n\right) \le \frac{2n+1}{2}.$$

This is also equivalent to

$$\sum_{k=1}^{2n} \left(\sqrt{k + n^2} - n\right) \le \frac{2n+1}{2}.$$

Since

$$\sqrt{k + n^2} - n = \frac{k + n^2 - n^2}{n + \sqrt{k + n^2}} < \frac{k}{2n},$$

we obtain

$$\sum_{k=1}^{2n} \left(\sqrt{k + n^2} - n\right) < \sum_{k=1}^{2n} \frac{k}{2n} = \frac{2n(2n+1)}{4n} = \frac{2n+1}{2},$$

which is exactly what we needed.

Solutions to introductory problems 155

50. Let a, b be real numbers. Solve in real numbers the system
$$\begin{cases} x + y = \sqrt[3]{a+b} \\ x^4 - y^4 = ax - by. \end{cases}$$

Solution. Write the system as
$$\begin{cases} (x+y)^3 = a+b \\ x^4 - y^4 = ax - by. \end{cases}$$

and consider it as a linear system in a and b. We express $b = (x+y)^3 - a$ from the first equation and insert this value in the second equation, obtaining therefore
$$x^4 - y^4 = ax - y(x+y)^3 + ay.$$

This can be also written as
$$a(x+y) = y(x+y)^3 + x^4 - y^4 = (x+y)(y(x+y)^2 + (x-y)(x^2+y^2)).$$

Suppose that $x + y = 0$, then coming back to the original system we obtain $a + b = 0$. Thus if $a + b \neq 0$, then no solution satisfies $x + y = 0$, while if $a + b = 0$, then all $(t, -t)$, with $t \in \mathbf{R}$, are solutions of the system.

Suppose now that $x + y \neq 0$, so that the previous relation becomes (after division by $x + y$)
$$a = y(x+y)^2 + (x-y)(x^2+y^2) = y(x^2+2xy+y^2) + x^3 + xy^2 - x^2y - y^3$$
$$= 3xy^2 + x^3.$$

Therefore
$$b = (x+y)^3 - a = (x+y)^3 - (3xy^2 + x^3) = y^3 + 3x^2y$$

and finally
$$a - b = x^3 + 3xy^2 - 3x^2y - y^3 = (x-y)^3.$$

Hence
$$\begin{cases} x + y = \sqrt[3]{a+b} \\ x - y = \sqrt[3]{a-b} \end{cases}$$

and we finally obtain the solution
$$x = \frac{\sqrt[3]{a+b} + \sqrt[3]{a-b}}{2}, \quad y = \frac{\sqrt[3]{a+b} - \sqrt[3]{a-b}}{2}.$$

51. Prove that for all $a, b, c > 0$ we have
$$\frac{abc}{a^3 + b^3 + abc} + \frac{abc}{b^3 + c^3 + abc} + \frac{abc}{c^3 + a^3 + abc} \leq 1.$$

Solution. We have
$$a^3 + b^3 = (a+b)(a^2 - ab + b^2) \geq (a+b)ab,$$
thus
$$a^3 + b^3 + abc \geq ab(a+b+c)$$
and
$$\frac{abc}{a^3 + b^3 + abc} \leq \frac{c}{a+b+c}.$$
Adding this inequality to two similar ones obtained by permuting cyclically a, b, c yields the desired result.

52. Define a sequence $(a_n)_{n \geq 1}$ by
$$a_1 = 1, \quad a_{n+1} = \frac{1 + 4a_n + \sqrt{1 + 24a_n}}{16}.$$
Find an explicit formula for a_n.

Solution. Let us simplify the problem a little bit, by introducing the sequence $b_n = \sqrt{1 + 24a_n}$. Thus $a_n = \frac{b_n^2 - 1}{24}$ and replacing this in the given relation yields
$$\frac{4}{6}(b_{n+1}^2 - 1) = 1 + \frac{b_n^2 - 1}{6} + b_n,$$
which in turn simplifies to
$$4b_{n+1}^2 = b_n^2 + 6b_n + 9 = (b_n + 3)^2.$$
Since $b_n > 0$ and $b_{n+1} > 0$, this is in turn equivalent to
$$2b_{n+1} = b_n + 3.$$
Now the problem is much easier, since we have to solve a linear recurrence of the first order. We look for α such that $2(b_{n+1} - \alpha) = b_n - \alpha$ for all n. The previous relation gives $\alpha = 3$. Now the sequence $(b_n - 3)_n$ becomes a geometric progression with ratio $1/2$ and first term
$$b_1 - 3 = \sqrt{1 + 24a_1} - 3 = 5 - 3 = 2,$$
thus
$$b_n - 3 = 2 \cdot 2^{1-n} \quad \text{so} \quad b_n = 3 + 2^{2-n}.$$
Finally, replacing this in $a_n = \frac{b_n^2 - 1}{24}$ gives the explicit expression
$$a_n = \frac{(3 + 2^{2-n})^2 - 1}{24}.$$

15 Solutions to advanced problems

1. If x, y are positive real numbers, define
$$x * y = \frac{x+y}{1+xy}.$$
Find $(...(((2*3)*4)*5)...)*1995$.

   Solution. The key point is to observe that
$$\frac{1+x*y}{1-x*y} = \frac{1+\frac{x+y}{1+xy}}{1-\frac{x+y}{1+xy}} =$$
$$\frac{(1+x)(1+y)}{(1-x)(1-y)} = \frac{1+x}{1-x} \cdot \frac{1+y}{1-y}.$$
   We deduce by an immediate induction that if
$$z = (...(((x_1 * x_2) * x_3)...) * x_k,$$
   then
$$\frac{1+z}{1-z} = \frac{1+x_1}{1-x_1} \cdot ... \cdot \frac{1+x_k}{1-x_k}.$$
   In particular, we have
$$\frac{1+(...(((2*3)*4)*5)...)*1995}{1-(...(((2*3)*4)*5)...)*1995} = \frac{1+2}{1-2} \cdot \frac{1+3}{1-3} \cdot ... \cdot \frac{1+1995}{1-1995} =$$
$$= \frac{3 \cdot 4 \cdot ... \cdot 1996}{(-1)^{1994} 1 \cdot 2 \cdot ... \cdot 1994} = \frac{1995 \cdot 1996}{2}.$$
   Solving the last equation yields
$$(...(((2*3)*4)*5)...)*1995 = \frac{1\,991\,009}{1\,991\,011}.$$

2. Three real numbers are given, such that the fractional part of the product of every two of them is $\frac{1}{2}$. Prove that these numbers are irrational.

   Solution. Let a, b, c be the given numbers. By hypothesis we can find integers x, y, z such that
$$bc = x + \frac{1}{2}, \quad ca = y + \frac{1}{2}, \quad ab = z + \frac{1}{2}.$$
   If we manage to prove that abc is irrational, then we are done, since then
$$a = \frac{abc}{bc} = \frac{abc}{x + \frac{1}{2}}$$

is irrational, and similarly for b and c. So suppose that abc is rational. Multiplying the previous relations yields

$$(abc)^2 = \left(x + \frac{1}{2}\right) \cdot \left(y + \frac{1}{2}\right) \cdot \left(z + \frac{1}{2}\right)$$

or

$$2(2abc)^2 = (2x+1)(2y+1)(2z+1).$$

Write $2abc = \frac{p}{q}$ for some relatively prime integers p, q, with $q \neq 0$. Then 2 divides $(2x+1)(2y+1)(2z+1)q^2$, thus 2 divides q^2. That means that q is even. But then 4 divides

$$(2x+1)(2y+1)(2z+1)q^2 = 2p^2,$$

thus 2 divides p^2 and then p is even. This contradicts the fact that p and q are relatively prime and finishes the proof of the fact that abc is irrational. The reader has probably already noticed that this problem is a variation on the standard proof that $\sqrt{2}$ is irrational.

3. Let a, b, c be real numbers satisfying

$$\begin{cases} (a+b)(b+c)(c+a) = abc \\ (a^3+b^3)(b^3+c^3)(c^3+a^3) = a^3b^3c^3. \end{cases}$$

Prove that $abc = 0$.

Solution. Let us argue by contradiction and suppose that none of the numbers a, b, c is zero. The second equation can be written

$$(a+b)(b+c)(c+a)(a^2-ab+b^2)(b^2-bc+c^2)(c^2-ca+a^2) = (abc)^3.$$

Combined with the first equation, after division by $abc \neq 0$ this yields

$$(a^2-ab+b^2)(b^2-bc+c^2)(c^2-ca+a^2) = a^2b^2c^2.$$

On the other hand, the AM-GM inequality yields

$$a^2 + b^2 \geq 2|ab| \geq ab + |ab|, \quad \text{hence} \quad a^2 - ab + b^2 \geq |ab|,$$

with equality if and only if $|a| = |b|$ and $ab = |ab|$, which means that $a = b$. Writing down two similar inequalities and multiplying them yields

$$(a^2-ab+b^2)(b^2-bc+c^2)(c^2-ca+a^2) \geq |ab| \cdot |bc| \cdot |ca| = a^2b^2c^2.$$

Since by hypothesis we have equality in the preceding inequality, we must have equality in all three inequalities written above. Thus $a = b = c$. But then the first equation gives $8a^3 = a^3$, hence $a = 0$, a contradiction. Thus the initial assumption was wrong and we have indeed $abc = 0$.

4. Real numbers a, b satisfy $a^3 - 3a^2 + 5a - 17 = 0$ and $b^3 - 3b^2 + 5b + 11 = 0$. Find $a + b$.

Solution. Let $x = a - 1$ and $y = b - 1$. Since

$$a^3 - 3a^2 + 5a - 17 = a^3 - 3a^2 + 3a - 1 + 2a - 16$$
$$= x^3 + 2(a - 1) - 14 = x^3 + 2x - 14,$$

we obtain $x^3 + 2x = 14$. Similarly, the second equation can be written $y^3 + 2y = -14$. Adding up these two relations yields

$$x^3 + y^3 + 2(x + y) = 0,$$

that is

$$(x + y)(x^2 - xy + y^2 + 2) = 0.$$

Since

$$x^2 - xy + y^2 + 2 = \left(x - \frac{y}{2}\right)^2 + \frac{3}{4}y^2 + 2 > 0,$$

we deduce that $x + y = 0$ and so $a + b = 2$. Another way to conclude once we have established the equality $x^3 + y^3 + 2(x + y) = 0$ is to note that the map $f(x) = x^3 + 2x$ is increasing on $\mathbf{R}$ (since it is the sum of two increasing functions) and that the relation can be written as $f(x) = f(-y)$. Thus necessarily $x = -y$ and $x + y = 0$.

5. Prove that

$$\sqrt{1 + \frac{1}{1^2} + \frac{1}{2^2}} + \sqrt{1 + \frac{1}{2^2} + \frac{1}{3^2}} + \ldots + \sqrt{1 + \frac{1}{1999^2} + \frac{1}{2000^2}}$$

is a rational number and compute it in lowest form.

Solution. Let us focus on the general term

$$\sqrt{1 + \frac{1}{n^2} + \frac{1}{(n+1)^2}} = \frac{\sqrt{n^2(n+1)^2 + n^2 + (n+1)^2}}{n(n+1)}.$$

On the other hand, expanding everything and rearranging terms yields

$$n^2(n+1)^2 + n^2 + (n+1)^2 = n^4 + 2n^3 + n^2 + n^2 + n^2 + 2n + 1 =$$
$$n^4 + 2n^3 + 3n^2 + 2n + 1 = (n^2 + n + 1)^2.$$

We conclude that

$$\sqrt{1 + \frac{1}{n^2} + \frac{1}{(n+1)^2}} = \frac{n^2 + n + 1}{n(n+1)} = 1 + \frac{1}{n(n+1)} = 1 + \frac{1}{n} - \frac{1}{n+1},$$

which already makes it clear that the number in the statement of the problem is rational. To explicitly compute it, note that we obtain a telescoping sum, which simplifies quite nicely:

$$\sqrt{1+\frac{1}{1^2}+\frac{1}{2^2}}+\sqrt{1+\frac{1}{2^2}+\frac{1}{3^2}}+\cdots+\sqrt{1+\frac{1}{1999^2}+\frac{1}{2000^2}}$$

$$=\left(1+\frac{1}{1}-\frac{1}{2}\right)+\left(1+\frac{1}{2}-\frac{1}{3}\right)+\cdots+\left(1+\frac{1}{1999}-\frac{1}{2000}\right)$$

$$=1999+\frac{1}{1}-\frac{1}{2000}=2000-\frac{1}{2000}=\frac{3999999}{2000}.$$

which is the answer of the problem.

6. Find a polynomial with integer coefficients having $\sqrt[5]{2+\sqrt{3}}+\sqrt[5]{2-\sqrt{3}}$ as a root.

   Solution. Let $a=\sqrt[5]{2+\sqrt{3}}$ and $b=\sqrt[5]{2-\sqrt{3}}$. The key relation is

   $$ab=\sqrt[5]{(2+\sqrt{3})(2-\sqrt{3})}=1.$$

   On the other hand, we have $a^5+b^5=4$. We need to find a polynomial with integer coefficients having $S=a+b$ as a root. We have

   $$4=a^5+b^5=(a+b)(a^4-a^3b+a^2b^2-ab^3+b^4)=S(a^4+b^4-(a^2+b^2)+1),$$

   since $ab=1$. On the other hand,

   $$a^2+b^2=S^2-2ab=S^2-2,$$

   and

   $$a^4+b^4=(a^2+b^2)^2-2a^2b^2=(S^2-2)^2-2=S^4-4S^2+2.$$

   Hence

   $$4=S(S^4-4S^2+2-S^2+2+1)$$

   and so

   $$x^5-5x^3+5x-4=0.$$

   This shows that the polynomial x^5-5x^3+5x-4 is a solution of the problem (one can actually prove that 5 is the least degree of a polynomial with integer coefficients having $\sqrt[5]{2+\sqrt{3}}+\sqrt[5]{2-\sqrt{3}}$ as a root).

7. Solve in real numbers the equation

   $$(3x+1)(4x+1)(6x+1)(12x+1)=5.$$

Solution. We will suitably multiply each factor in the left hand-side, so that after this operation we reduce the equation to some quadratic equations. In order to do that, we multiply $3x+1$ by 4, $4x+1$ by 3 and $6x+1$ by 2. The new (and equivalent) equation becomes

$$(12x+4)(12x+3)(12x+2)(12x+1) = 120.$$

Set $t = 12x$, so that

$$(t+1)(t+2)(t+3)(t+4) = 120.$$

Pairing the first and the last factor (and similarly the second and the third one) gives the equivalent equation

$$(t^2 + 5t + 4)(t^2 + 5t + 6) = 120.$$

This reduces to the quadratic equation $y(y+2) = 120$ after the substitution $y = t^2 + 5t + 4$. This equation has the solutions $y = -12$ and $y = 10$. Now the problem is reduced to finding t for which $t^2 + 5t + 16 = 0$ or $t^2 + 5t - 6 = 0$. The first equation has no solutions, since the discriminant is negative. The second one has the solutions $t = -6$ and $t = 1$. Taking into account that $t = 12x$, we finally obtain the solutions $x = \frac{1}{12}, -\frac{1}{2}$.

8. Let n be a positive integer. Solve in real numbers the equation

$$\lfloor x \rfloor + \lfloor 2x \rfloor + ... + \lfloor nx \rfloor = \frac{n(n+1)}{2}.$$

Solution. Let x be a solution of the equation. Then certainly

$$\frac{n(n+1)}{2} = \lfloor x \rfloor + \lfloor 2x \rfloor + ... + \lfloor nx \rfloor \leq x + 2x + ... + nx = \frac{n(n+1)}{2}x,$$

thus $x \geq 1$. Let us write $x = 1 + y$, with $y \geq 0$. Then

$$\lfloor kx \rfloor = k + \lfloor ky \rfloor,$$

so the equation can be written

$$1 + \lfloor y \rfloor + 2 + \lfloor 2y \rfloor + ... + n + \lfloor ny \rfloor = \frac{n(n+1)}{2},$$

that is

$$\lfloor y \rfloor + \lfloor 2y \rfloor + ... + \lfloor ny \rfloor = 0.$$

We are now in pretty good shape, since each of the numbers $\lfloor y \rfloor, ..., \lfloor ny \rfloor$ is nonnegative (because y is so). Thus the previous relation holds if and only if all numbers $\lfloor y \rfloor, ..., \lfloor ny \rfloor$ are 0. This happens if and only if $y < \frac{1}{n}$, giving the solution $x \in [1, 1 + \frac{1}{n})$.

9. Any two of the real numbers a_1, a_2, a_3, a_4, a_5 differ by no less than 1. Moreover, there is a real number k satisfying
$$\begin{cases} a_1 + a_2 + a_3 + a_4 + a_5 = 2k \\ a_1^2 + a_2^2 + a_3^2 + a_4^2 + a_5^2 = 2k^2. \end{cases}$$
Prove that $k^2 \geq \frac{25}{3}$.

Solution. By symmetry, we may assume that $a_5 > a_4 > \cdots > a_1$. Combined with the hypothesis, we obtain $a_{i+1} - a_i \geq 1$ for all i, hence
$$a_i - a_j = a_i - a_{i-1} + a_{i-1} - a_{i-2} + \ldots + a_{j+1} - a_j \geq i - j$$
for all $5 \geq i > j \geq 1$. It follows that
$$\sum_{1 \leq j < i \leq 5} (a_i - a_j)^2 \geq \sum_{1 \leq j < i \leq 5} (i-j)^2 = 4 \cdot 1^2 + 3 \cdot 2^2 + 2 \cdot 3^2 + 4^2 = 50.$$
This can also be written as
$$4 \sum_{i=1}^{5} a_i^2 - 2 \sum_{1 \leq j < i \leq 5} a_i a_j \geq 50.$$
Combined with the equality
$$\sum_{i=1}^{5} a_i^2 + 2 \sum_{1 \leq j < i \leq 5} a_i a_j = (\sum_{i=1}^{5} a_i)^2 = 4k^2$$
and with the hypothesis, this yields
$$10k^2 = 5 \sum_{i=1}^{5} a_i^2 \geq 50 + 4k^2.$$
Thus $k^2 \geq \frac{25}{3}$, which is the desired result.

10. Prove that if a, b, c, d are nonzero real numbers, not all equal, and if
$$a + \frac{1}{b} = b + \frac{1}{c} = c + \frac{1}{d} = d + \frac{1}{a},$$
then $|abcd| = 1$.

Solution. Write the first equation as
$$a - b = \frac{1}{c} - \frac{1}{b} = \frac{b-c}{bc}.$$

We do the same with the other equalities, obtaining
$$b - c = \frac{c-d}{cd}, \quad c - d = \frac{d-a}{da}, \quad d - a = \frac{a-b}{ab}.$$
Now we multiply all these equalities and obtain
$$(a-b)(b-c)(c-d)(d-a) = \frac{(a-b)(b-c)(c-d)(d-a)}{a^2b^2c^2d^2}.$$
If $(a-b)(b-c)(c-d)(d-a)$ was nonzero, then the previous relation would yield $(abcd)^2 = 1$, thus $|abcd| = 1$. Let us suppose that $(a-b)(b-c)(c-d)(d-a) = 0$ and, for example, $a = b$. Then $a - b = \frac{b-c}{bc}$ yields $b = c$. Similarly $c = d$ and so $a = b = c = d$, contradicting the hypothesis. Thus $(a-b)(b-c)(c-d)(d-a) \neq 0$ and the problem is solved.

11. The side lengths of a triangle are the roots of a cubic polynomial with rational coefficients. Prove that the altitudes of this triangle are roots of a polynomial of sixth degree with rational coefficients.

    Solution. Let h_a be the altitude corresponding to the side with length a. The key formula here is $h_a = \frac{2S}{a}$, where S is the area of the triangle. On the other hand, by Heron's formula we have
    $$16S^2 = (a+b+c)(b+c-a)(c+a-b)(a+b-c).$$
    Let $f(x) = x^3 + Ax^2 + Bx + C$ be a polynomial with rational coefficients having roots a, b, c. Then $a + b + c = -A$ by Vieta's formulae, so
    $$(b+c-a)(c+a-b)(a+b-c) = (-A-2a)(-A-2b)(-A-2c)$$
    $$= 8f\left(-\frac{A}{2}\right) \in \mathbf{Q}$$
    We conclude that $S^2 \in \mathbf{Q}$. On the other hand, we have $f(a) = 0$, so $f\left(\frac{2S}{h_a}\right) = 0$, which can be written as
    $$8S^3 + 4S^2 A h_a + 2SB h_a^2 + C h_a^3 = 0.$$
    Write this as
    $$2S(4S^2 + Bh_a^2) = -(4S^2 A h_a + C h_a^3)$$
    and square it. We obtain that h_a is a root of the polynomial
    $$G(x) = (4S^2 Ax + Cx^3)^2 - 4S^2(4S^2 + x^2 B)^2,$$
    which has rational coefficients because $S^2, A, B, C \in \mathbf{Q}$. Moreover, by symmetry G also vanishes at h_b and h_c, and the problem is solved (note that G is nonzero, since the coefficient of x^6 is $C^2 \neq 0$).

12. Find the maximal value of
$$\frac{(1+x)^8 + 16x^4}{(1+x^2)^4}$$
when $x \in \mathbf{R}$.

Solution. The key idea is to observe that we can rewrite the expression in a much simpler way:
$$\frac{(1+x)^8 + 16x^4}{(1+x^2)^4} = \frac{(1+x)^8}{(1+x^2)^4} + \frac{(2x)^4}{(1+x^2)^4} =$$
$$\left(\frac{(1+x)^2}{1+x^2}\right)^4 + \left(\frac{2x}{1+x^2}\right)^4 = \left(1 + \frac{2x}{1+x^2}\right)^4 + \left(\frac{2x}{1+x^2}\right)^4.$$

Letting $t(x) = \frac{2x}{1+x^2}$, the problem comes down to finding the maximal value of $(1+t(x))^4 + t(x)^4$ when x ranges over the real numbers. Let us start with the easier question of finding the maximal value of $t(x)$. This is rather simple, since
$$t(x) = \frac{2x}{1+x^2} \leq 1, \quad \text{as} \quad x^2 + 1 \geq 2x.$$

On the other hand, $t(1) = 1$, so the maximal value of t is 1. It follows that the maximal value of $(1+t(x))^4 + t(x)^4$ is $2^4 + 1 = 17$, obtained when $x = 1$. Hence the answer of the problem is 17.

13. Let x, y, z be real numbers greater than -1. Prove that
$$\frac{1+x^2}{1+y+z^2} + \frac{1+y^2}{1+z+x^2} + \frac{1+z^2}{1+x+y^2} \geq 2.$$

Solution. Note first that the denominators are positive by hypothesis. The key point is to bound y from above by an expression involving only y^2. This is easily achieved using the AM-GM inequality, since $y \leq \frac{1+y^2}{2}$. Thus
$$\frac{1+x^2}{1+y+z^2} \geq \frac{1+x^2}{1+z^2+\frac{1+y^2}{2}} = \frac{2(1+x^2)}{2(1+z^2)+(1+y^2)}.$$

Taking $a = 1+x^2$, $b = 1+y^2$ and $c = 1+z^2$ as new variables, it suffices to prove the inequality
$$\frac{a}{2c+b} + \frac{b}{2a+c} + \frac{c}{2b+a} \geq 1.$$

This is an easy consequence of the Cauchy-Schwarz inequality:

$$\frac{a^2}{2ac+ab} + \frac{b^2}{2ab+bc} + \frac{c^2}{2bc+ac} \geq \frac{(a+b+c)^2}{3(ab+bc+ca)} \geq 1,$$

the last inequality being equivalent (after expansion) to $a^2 + b^2 + c^2 \geq ab + bc + ca$, or $(a-b)^2 + (b-c)^2 + (c-a)^2 \geq 0$.

14. Let a and b be real numbers. Prove that

$$a^3 + b^3 + (a+b)^3 + 6ab = 16$$

if and only if $a + b = 2$.

Solution. Let us denote $s = a + b$ and $p = ab$. Then

$$a^3 + b^3 = (a+b)(a^2 - ab + b^2) = s((a+b)^2 - 3ab) = s(s^2 - 3p) = s^3 - 3ps.$$

The relation $a^3 + b^3 + (a+b)^3 + 6ab = 16$ is equivalent to

$$s^3 - 3sp + s^3 + 6p = 16,$$

or equivalently $2(s^3 - 8) = 3p(s - 2)$. The good news is that both factors are multiples of $s - 2$ so the previous relation is equivalent to

$$(s-2)[2(s^2 + 2s + 4) - 3p] = 0.$$

If we manage to prove that $2(s^2 + 2s + 4) \neq 3p$, then we are done. Suppose by contradiction that $2(s^2 + 2s + 4) = 3p$, that is

$$2(a+b)^2 + 4(a+b) + 8 = 3ab.$$

Expanding yields the equivalent form

$$2a^2 + 2b^2 + ab + 4a + 4b + 8 = 0.$$

We can rewrite this as follows:

$$a^2 + ab + b^2 + a^2 + 4a + 4 + b^2 + 4b + 4 = 0,$$

that is

$$a^2 + ab + b^2 + (a+2)^2 + (b+2)^2 = 0.$$

But

$$a^2 + ab + b^2 = a^2 + ab + \frac{1}{4}b^2 + \frac{3}{4}b^2 = \left(a + \frac{b}{2}\right)^2 + \frac{3}{4}b^2 \geq 0,$$

hence $a^2 + ab + b^2 + (a+2)^2 + (b+2)^2 = 0$ forces $a = -2$, $b = -2$ and $a^2 + ab + b^2 = 0$. This is clearly impossible, hence the result follows.

15. If a and b are nonzero real numbers such that

$$20a + 21b = \frac{a}{a^2 + b^2} \quad \text{and} \quad 21a - 20b = \frac{b}{a^2 + b^2},$$

evaluate $a^2 + b^2$.

Solution. This is a quite tricky application of Lagrange's identity:

$$(20a + 21b)^2 + (21a - 20b)^2 = (20^2 + 21^2)(a^2 + b^2).$$

The right hand-side equals $841(a^2 + b^2)$, while by hypothesis the left hand-side equals

$$\frac{a^2}{(a^2+b^2)^2} + \frac{b^2}{(a^2+b^2)^2} = \frac{a^2+b^2}{(a^2+b^2)^2} = \frac{1}{a^2+b^2}.$$

Thus

$$\frac{1}{a^2+b^2} = 841(a^2+b^2)$$

and since $a^2 + b^2$ is positive, we conclude that

$$a^2 + b^2 = \sqrt{\frac{1}{841}} = \frac{1}{29}.$$

16. Let a, b be solutions of the equation $x^4 + x^3 - 1 = 0$. Prove that ab is a solution of the equation $x^6 + x^4 + x^3 - x^2 - 1 = 0$.

Solution. Since a and b are roots of $x^4 + x^3 - 1$, we can factor to get

$$x^4 + x^3 - 1 = (x-a)(x-b)(x^2 - wx + t)$$

for some real numbers w and t. Introducing new variables $u = a + b$ and $v = ab$, we can expand this and match coefficients to get

$$u + w = -1, \quad uw + v + t = 0, \quad vw + ut = 0, \quad \text{and} \quad vt = -1.$$

We need to eliminate u, w, t among these equations and prove that v is a solution of $x^6 + x^4 + x^3 - x^2 - 1 = 0$. Elimination of variables is quite easy: we express $t = -\frac{1}{v}$ from the last relation, then $w = \frac{u}{v^2}$ from the third relation. Next, we replace these values in the first two relations and get

$$u + \frac{u}{v^2} = -1, \quad \frac{u^2}{v^2} + v - \frac{1}{v} = 0.$$

We finally replace $u = -\frac{v^2}{v^2+1}$ in the second relation. A small computation yields the desired result.

Here is an alternative way to start the solution: if c, d are the other two solutions of the equation, then Vieta's relations yield

$$a + b + c + d = -1, \quad ab + bc + cd + da + ac + bd = 0,$$

$$abc + bcd + cda + dab = 0, \quad abcd = -1.$$

Writing the third relation as

$$ab(c + d) + cd(a + b) = 0,$$

we introduce the new variables $u = a + b$, $v = ab$, $w = c + d$ and $t = cd$. Since

$$ab+bc+cd+da+ac+bd = ad+ac+bc+bd+ab+cd = (a+b)(c+d)+ab+cd,$$

we can write the previous relations purely in terms of u, v, w, t as

$$u + w = -1, \quad uw + v + t = 0, \quad vw + ut = 0, \quad vt = -1.$$

17. Find all real numbers x such that

$$\sqrt{x + 2\sqrt{x + 2\sqrt{x + 2\sqrt{3x}}}} = x.$$

Solution. We start by looking for integer solutions. Trial and error gives the solution $x = 3$. We claim that this is the unique solution of the equation. Dividing by x, we obtain the equivalent equation

$$\sqrt{\frac{1}{x} + 2\sqrt{\frac{1}{x^3} + 2\sqrt{\frac{1}{x^7} + 2\sqrt{\frac{3}{x^{15}}}}}} = 1.$$

The left hand-side is a decreasing function of x, hence the equation has at most one positive solution. Since we have already found one, we are done: the only real number satisfying the original equation is $x = 3$.

Here is an alternative solution, based on Corollary 9.2. The left hand side is not defined if $x < 0$, hence we may assume $x \geq 0$. For fixed $x \geq 0$, define $f(t) = \sqrt{x + 2t}$. It is easy to see that $f(t)$ is an increasing function of t for $t \geq 0$. Hence by Corollary 9.2 the solutions to $f(f(f(f(t)))) = t$ with $t \geq 0$ are just the solutions to $f(t) = t$. However, the original equation just says that $f(f(f(f(x)))) = x$. Hence we only need to solve $f(x) = x$ or $\sqrt{3x} = x$. This is easily seen to give $x = 0$ or $x = 3$.

18. Let a_1, a_2, a_3, a_4, a_5 be real numbers satisfying
$$\frac{a_1}{k^2+1} + \frac{a_2}{k^2+2} + \frac{a_3}{k^2+3} + \frac{a_4}{k^2+4} + \frac{a_5}{k^2+5} = \frac{1}{k^2}$$
for $1 \le k \le 5$. Find the value of $\frac{a_1}{37} + \frac{a_2}{38} + \frac{a_3}{39} + \frac{a_4}{40} + \frac{a_5}{41}$.

Solution. This is a linear system in $a_1, ..., a_5$, so with enough patience one should in principle be able to solve it and find $a_1, ..., a_5$, then compute the desired expression. It should be clear that this is not what we are looking for! Let us consider
$$F(x) = \frac{a_1}{x+1} + \frac{a_2}{x+2} + ... + \frac{a_5}{x+5} - \frac{1}{x}.$$
By hypothesis, F vanishes at $1^2, ..., 5^2$. On the other hand, by clearing denominators we can certainly write
$$F(x) = \frac{P(x)}{x(x+1)...(x+5)}$$
for some polynomial P of degree at most 5. Now P vanishes at $1^2, ..., 5^2$, which are distinct numbers, and P has degree at most 5. Hence there must be a constant c such that
$$P(x) = c(x-1^2)(x-2^2)...(x-5^2).$$
How to find c? Observe that
$$\frac{P(x)}{(x+1)...(x+5)} = xF(x) = \frac{xa_1}{x+1} + ... + \frac{xa_5}{x+1} - 1$$
and so
$$\frac{P(0)}{5!} = -1.$$
Thus $-c \cdot 5!^2 = -5!$ and $c = \frac{1}{5!}$. Now, the desired expression
$$\frac{a_1}{37} + \frac{a_2}{38} + \frac{a_3}{39} + \frac{a_4}{40} + \frac{a_5}{41} = \frac{a_1}{6^2+1} + ... + \frac{a_5}{6^2+5} =$$
$$\frac{1}{6^2} + \frac{P(6^2)}{6^2(6^2+1)...(6^2+5)} = \frac{1}{36} + \frac{(6^2-1)...(6^2-5^2)}{5! \cdot (6^2+1)...(6^2+5)}.$$
We can simplify a little bit the previous expression:
$$\frac{(6^2-1)...(6^2-5^2)}{5! \cdot (6^2+1)...(6^2+5)} = \frac{(6-1)...(6-5) \cdot 7 \cdot 8 \cdot 9 \cdot 10 \cdot 11}{5! \cdot 36 \cdot 37 \cdot ... \cdot 41} =$$
$$\frac{7 \cdot 8 \cdot 9 \cdot 10 \cdot 11}{36 \cdot 37 \cdot 38 \cdot 39 \cdot 40 \cdot 41} = \frac{7 \cdot 11}{12 \cdot 37 \cdot 19 \cdot 13 \cdot 41}.$$

19. Are there nonzero real numbers a, b, c such that for all $n > 3$ there is a polynomial $P_n(x) = x^n + \cdots + ax^2 + bx + c$ which has exactly n (not necessary distinct) integral roots?

Solution. The answer is negative. Suppose that $r_1, ..., r_n$ are all roots of P_n and that they are all integers. Then by Vieta's relations

$$r_1 r_2 ... r_n = (-1)^n c, \quad r_2 r_3 ... r_n + ... + r_1 ... r_{n-1} = (-1)^{n-1} b$$

and

$$r_3 ... r_n + ... + r_1 ... r_{n-2} = (-1)^{n-2} a.$$

We write the second and third relations as

$$\frac{1}{r_1} + ... + \frac{1}{r_n} = -\frac{b}{c}, \quad \sum_{i<j} \frac{1}{r_i r_j} = \frac{a}{c},$$

taking into account the first relation. Combined with the identity

$$(x_1 + ... + x_n)^2 = x_1^2 + ... + x_n^2 + 2 \sum_{i<j} x_i x_j$$

this yields

$$\sum_{i=1}^{n} \frac{1}{r_i^2} = \frac{b^2}{c^2} - 2\frac{a}{c}.$$

On the other hand the AM-GM inequality gives

$$\sum_{i=1}^{n} \frac{1}{r_i^2} \geq \frac{n}{\sqrt[n]{r_1^2 ... r_n^2}} = \frac{n}{\sqrt[n]{c^2}}.$$

Hence if $u = \frac{b^2}{c^2} - 2\frac{a}{c}$, then $u \sqrt[n]{c^2} \geq n$ for all $n > 3$, in particular $n \leq u \cdot |c|$ for all $n > 3$. Since this is clearly absurd, the result follows. Note that we only used the hypothesis that the roots are real, not the fact that they are integers.

20. Let $n > 1$ be an integer and let $a_0, a_1, ..., a_n$ be real numbers such that $a_0 = \frac{1}{2}$ and

$$a_{k+1} = a_k + \frac{a_k^2}{n} \quad \text{for} \quad k = 0, 1, ... n - 1.$$

Prove that $1 - \frac{1}{n} < a_n < 1$.

Solution. The key point is to observe that

$$\frac{1}{a_{k+1}} = \frac{1}{a_k + \frac{a_k^2}{n}} = \frac{n}{a_k(n + a_k)} = \frac{1}{a_k} - \frac{1}{n + a_k}.$$

Adding these relations for $k = 0, 1, ..., n - 1$, we obtain a telescoping sum, which simplifies to

$$\frac{1}{a_n} = \frac{1}{a_0} - \sum_{k=0}^{n-1} \frac{1}{n + a_k} = 2 - \sum_{k=0}^{n-1} \frac{1}{n + a_k}.$$

Next, it is clear from the relation defining $a_0, ..., a_n$ that the sequence is increasing, hence

$$\frac{1}{n} > \frac{1}{n + a_k} > \frac{1}{n + a_n}$$

for $0 \leq k < n$. Adding up these relations yields

$$1 > \sum_{k=0}^{n-1} \frac{1}{n + a_k} > \frac{n}{n + a_n}.$$

Combining these inequalities with the relation

$$\sum_{k=0}^{n-1} \frac{1}{n + a_k} = 2 - \frac{1}{a_n}$$

established above, we obtain

$$1 > 2 - \frac{1}{a_n} > \frac{n}{n + a_n}.$$

The inequality on the left yields $a_n < 1$, so the inequality on the right can be weakened to

$$2 - \frac{1}{a_n} > \frac{n}{n + 1}.$$

This finally yields

$$a_n > \frac{n+1}{n+2} = 1 - \frac{1}{n+2} > 1 - \frac{1}{n}.$$

21. Let $x_i = \frac{i}{101}$. Compute

$$\sum_{i=0}^{101} \frac{x_i^3}{1 - 3x_i + 3x_i^2}.$$

Solution. The key point is to observe that $1 - 3x + 3x^2$ is a big part in the expansion of $(1-x)^3$. More precisely, we have

$$1 - 3x + 3x^2 = (1-x)^3 + x^3.$$

On the other hand,
$$1 - x_i = \frac{101 - i}{101} = x_{101-i}.$$

Hence
$$\sum_{i=0}^{101} \frac{x_i^3}{1 - 3x_i + 3x_i^2} = \sum_{i=0}^{101} \frac{x_i^3}{x_i^3 + x_{101-i}^3}.$$

Now, we split the sum in two parts:
$$\sum_{i=0}^{101} \frac{x_i^3}{x_i^3 + x_{101-i}^3} = \sum_{i=0}^{50} \frac{x_i^3}{x_i^3 + x_{101-i}^3} + \sum_{i=51}^{101} \frac{x_i^3}{x_i^3 + x_{101-i}^3}$$

and we make the change of variable $j = 101 - i$ in the second sum. We obtain

$$\sum_{i=0}^{50} \frac{x_i^3}{x_i^3 + x_{101-i}^3} + \sum_{i=51}^{101} \frac{x_i^3}{x_i^3 + x_{101-i}^3} = \sum_{i=0}^{50} \frac{x_i^3}{x_i^3 + x_{101-i}^3} + \sum_{j=0}^{50} \frac{x_{101-j}^3}{x_j^3 + x_{101-j}^3} =$$

$$\sum_{i=0}^{50} \frac{x_i^3 + x_{101-i}^3}{x_i^3 + x_{101-i}^3} = \sum_{i=0}^{50} 1 = 51.$$

Hence the result is 51.

22. Let a, b be real numbers and let $f(x) = x^2 + ax + b$. Suppose that the equation $f(f(x)) = 0$ has four different real solutions, and that the sum of two of these solutions is -1. Prove that $b \leq -\frac{1}{4}$.

    Solution. First, we claim that the equation $f(x) = 0$ has two distinct real roots x_1, x_2. If r is a real solution of the equation $f(f(x)) = 0$, then $f(r)$ is a real solution of the equation $f(x) = 0$. It remains to see that the equation $f(x) = 0$ does not have equal solutions. If x_1 was a double solution of the equation $f(x) = 0$, then all solutions r of the equation $f(f(x)) = 0$ would satisfy $f(r) = x_1$, and so the quadratic equation $f(t) = x_1$ would have four distinct solutions, a contradiction. This proves the claim.

    Now, the set of roots of the equation $f(f(x)) = 0$ is the union of the sets of roots of the equations $f(x) = x_1$ and $f(x) = x_2$. Let y_1, y_2 be two roots of the equation $f(f(x)) = 0$ with $y_1 + y_2 = -1$. If y_1, y_2 are both roots of one of the equations $f(x) = x_1$ or $f(x) = x_2$, say of the first one, then by Vieta's relations we would obtain $a = 1$. Next, since each of the equations $f(x) = x_1$ and $f(x) = x_2$ has real solutions, the discriminants of the corresponding equations are nonnegative, hence $\frac{1}{4} \geq b - x_1$ and

$\frac{1}{4} \geq b - x_2$. Adding these two relations and taking into account that $x_1 + x_2 = -1$, we obtain $b \leq -\frac{1}{4}$, and we are done.

So suppose that y_1 is a solution of $f(x) = x_1$ and y_2 is a solution of $f(x) = x_2$. Then

$$f(y_1) + f(y_2) = x_1 + x_2 = -a,$$

again by Vieta's relations. But since $y_1 + y_2 = -1$, we have

$$f(y_1) + f(y_2) = y_1^2 + y_2^2 - a + 2b.$$

We deduce that

$$b = -\frac{y_1^2 + y_2^2}{2}$$

and it remains to see that this is at most $-\frac{1}{4}$. Equivalently, we need to prove that $y_1^2 + y_2^2 \geq \frac{1}{2}$, which is clear, since

$$y_1^2 + y_2^2 \geq \frac{(y_1 + y_2)^2}{2} = \frac{1}{2}.$$

23. Real numbers a, b, c satisfy

$$\frac{a}{a^2 - bc} + \frac{b}{b^2 - ca} + \frac{c}{c^2 - ab} = 0.$$

Prove that

$$\frac{a}{(a^2 - bc)^2} + \frac{b}{(b^2 - ca)^2} + \frac{c}{(c^2 - ab)^2} = 0.$$

Solution. Let us multiply the given relation by $\frac{1}{a^2 - bc}$. We obtain

$$\frac{a}{(a^2 - bc)^2} + \frac{b}{(a^2 - bc)(b^2 - ca)} + \frac{c}{(c^2 - ab)(a^2 - bc)} = 0.$$

We write down two similar equalities, obtained by multiplying the given relation by $\frac{1}{b^2 - ca}$ and $\frac{1}{c^2 - ab}$. Adding up these three relations yields

$$\frac{a}{(a^2 - bc)^2} + \frac{b}{(b^2 - ca)^2} + \frac{c}{(c^2 - ab)^2} + S = 0,$$

where

$$S = \frac{a}{(a^2 - bc)(b^2 - ca)} + \frac{a}{(a^2 - bc)(c^2 - ab)} + \frac{b}{(b^2 - ca)(c^2 - ab)} +$$

Solutions to advanced problems

$$+\frac{b}{(b^2-ac)(a^2-bc)}+\frac{c}{(c^2-ab)(a^2-bc)}+\frac{c}{(c^2-ab)(b^2-ac)}.$$

Clearing denominators, we obtain

$$S = \frac{a(c^2-ab+b^2-ac)+b(a^2-bc+c^2-ab)+c(a^2-bc+b^2-ca)}{(a^2-bc)(b^2-ca)(c^2-ab)}.$$

Finally,

$$a(c^2-ab+b^2-ac)+b(a^2-bc+c^2-ab)+c(a^2-bc+b^2-ca) =$$
$$a(c^2+b^2)+b(a^2+c^2)+c(a^2+b^2)-a^2(b+c)-b^2(c+a)-c^2(a+b) = 0,$$

thus $S = 0$. Coming back to the relation

$$\frac{a}{(a^2-bc)^2}+\frac{b}{(b^2-ca)^2}+\frac{c}{(c^2-ab)^2}+S = 0,$$

the result follows.

Note that during the proof we actually established the following curious identity, which holds for all real numbers a, b, c such that $a^2 \neq bc$, $b^2 \neq ca$ and $c^2 \neq ab$:

$$\left(\frac{1}{a^2-bc}+\frac{1}{b^2-ca}+\frac{1}{c^2-ab}\right)\cdot\left(\frac{a}{a^2-bc}+\frac{b}{b^2-ca}+\frac{c}{c^2-ab}\right) =$$
$$\frac{a}{(a^2-bc)^2}+\frac{b}{(b^2-ca)^2}+\frac{c}{(c^2-ab)^2}.$$

Here is an alternative approach, which describes all triples (a, b, c) satisfying the hypothesis of the problem. Note that if $a+b+c = 0$, then

$$a^2-bc = b^2-ac = c^2-ab = a^2+ab+b^2.$$

Therefore in this case

$$\frac{a}{(a^2-bc)^m}+\frac{b}{(b^2-ca)^m}+\frac{c}{(c^2-ab)^m} = \frac{a+b+c}{(a^2+ab+b^2)^m} = 0$$

for any m. In particular,

$$\frac{a}{a^2-bc}+\frac{b}{b^2-ca}+\frac{c}{c^2-ab}$$

is a multiple of $a+b+c$. Putting it over a common denominator and factoring gives (after some tedious algebra left to the reader)

$$\frac{a}{a^2-bc}+\frac{b}{b^2-ca}+\frac{c}{c^2-ab}$$
$$= -\frac{(a+b+c)(a^2(b-c)^2+b^2(c-a)^2+c^2(a-b)^2)}{(a^2-bc)(b^2-ca)(c^2-ab)}.$$

The second factor in the numerator can only vanish if $a = b = c$, but in this case all three denominators vanish. Hence we must exclude this case. Thus
$$\frac{a}{a^2 - bc} + \frac{b}{b^2 - ca} + \frac{c}{c^2 - ab} = 0$$
if and only if $a + b + c = 0$ and in this case
$$\frac{a}{(a^2 - bc)^2} + \frac{b}{(b^2 - ca)^2} + \frac{c}{(c^2 - ab)^2} = 0$$
as well.

24. Let n be a positive integer and let $a_k = 2^{2^{k-n}} + k$. Prove that
$$(a_1 - a_0)(a_2 - a_1)\ldots(a_n - a_{n-1}) = \frac{7}{a_0 + a_1}.$$

Solution. To simplify notations, we let $x = 2^{2^{-n}}$, so that $a_k = x^{2^k} + k$ and
$$a_k - a_{k-1} = x^{2^k} - x^{2^{k-1}} + 1.$$

We use the factorization
$$a^4 + a^2 + 1 = (a^2 - a + 1)(a^2 + a + 1)$$
to write
$$x^{2^k} - x^{2^{k-1}} + 1 = \frac{x^{2^{k+1}} + x^{2^k} + 1}{x^{2^k} + x^{2^{k-1}} + 1}.$$

Thus
$$(a_1 - a_0)(a_2 - a_1)\ldots(a_n - a_{n-1}) = \prod_{k=1}^{n}(a_k - a_{k-1}) =$$
$$\prod_{k=1}^{n} \frac{x^{2^{k+1}} + x^{2^k} + 1}{x^{2^k} + x^{2^{k-1}} + 1}.$$

The last product telescopes to
$$\frac{x^{2^{n+1}} + x^{2^n} + 1}{x^2 + x + 1}.$$

Now, let us not forget that $x = 2^{2^{-n}}$, hence $x^{2^n} = 2$ and $x^{2^{n+1}} = 4$. Thus
$$(a_1 - a_0)(a_2 - a_1)\ldots(a_n - a_{n-1}) = \frac{7}{x^2 + x + 1} = \frac{7}{a_0 + a_1},$$
since $a_0 = x$ and $a_1 = 1 + x^2$. The result follows.

25. Find the integer part of

$$1 + \frac{1}{\sqrt[3]{2^2}} + \frac{1}{\sqrt[3]{3^2}} + \ldots + \frac{1}{\sqrt[3]{(10^9)^2}}.$$

Solution. Let S denote the expression whose integer part we are looking for. The key idea is to find a good estimate of $\frac{1}{\sqrt[3]{n^2}}$ of the form

$$a_n - a_{n-1} \leq \frac{1}{\sqrt[3]{n^2}} \leq b_n - b_{n-1}$$

for some easy-to-compute sequences. If we manage to do this, then adding up the relations and using the fact that the sums are telescopic, we will find a good estimate on the desired sum. Now, let us find a_n and b_n. The key relation is

$$a - b = (\sqrt[3]{a} - \sqrt[3]{b})(\sqrt[3]{a^2} + \sqrt[3]{ab} + \sqrt[3]{b^2}).$$

This shows that if $a > b > 0$, then

$$\frac{\sqrt[3]{a} - \sqrt[3]{b}}{a - b} = \frac{1}{\sqrt[3]{a^2} + \sqrt[3]{ab} + \sqrt[3]{b^2}} \in \left(\frac{1}{3\sqrt[3]{a^2}}, \frac{1}{3\sqrt[3]{b^2}}\right),$$

that is

$$\frac{1}{3\sqrt[3]{a^2}} < \frac{\sqrt[3]{a} - \sqrt[3]{b}}{a - b} < \frac{1}{3\sqrt[3]{b^2}}.$$

Taking $a = n+1$ and $b = n$ yields

$$\frac{1}{3\sqrt[3]{(n+1)^2}} < \sqrt[3]{n+1} - \sqrt[3]{n} < \frac{1}{3\sqrt[3]{n^2}}.$$

Adding the relations on the left for $n = 1, \ldots, 10^9 - 1$ yields

$$S - 1 < 3(10^3 - 1) = 2997,$$

that is $S < 2998$. On the other hand, adding the inequalities on the right for $n = 1, \ldots, 10^9$ gives

$$3(10^3 - 1) < 3(\sqrt[3]{10^9 + 1} - 1) < S,$$

that is $S > 2997$. We conclude that $\lfloor S \rfloor = 2997$.

26. Is there a sequence $(a_n)_{n \geq 1}$ of positive real numbers such that

$$a_1 + a_2 + \ldots + a_n \leq n^2 \quad \text{and} \quad \frac{1}{a_1} + \frac{1}{a_2} + \ldots + \frac{1}{a_n} \leq 2008$$

for all positive integers n?

Solution. The answer is negative. Suppose by contradiction that such a sequence exists. Using the AM-HM inequality, we obtain for all positive integers n the inequality

$$\frac{1}{a_{n+1}} + \ldots + \frac{1}{a_{2n}} \geq \frac{n^2}{a_{n+1} + \ldots + a_{2n}} \geq \frac{n^2}{a_1 + \ldots + a_{2n}} \geq \frac{1}{4}.$$

Let

$$S_n = \frac{1}{a_1} + \frac{1}{a_2} + \ldots + \frac{1}{a_n}.$$

Then by our assumption $S_n \leq 2008$ for all $n \geq 1$, and by the previous discussion we have

$$S_{2n} - S_n \geq \frac{1}{4}$$

for all n. Hence

$$S_{2^{k+1}} - S_{2^k} \geq \frac{1}{4}$$

for all $k \geq 0$. Adding these relations for $k = 0, 1, \ldots, n-1$, we obtain for all n

$$2008 \geq S_{2^n} - S_1 \geq \frac{n}{4}.$$

This plain contradiction shows that our initial assumption was wrong, hence such a sequence does not exist.

27. Is there a polynomial f with integer coefficients such that $f(x, y, z)$ and $x + \sqrt[3]{2}y + \sqrt[3]{3}z$ have the same sign for all integers x, y, z?

Solution. The answer is positive, but finding such a polynomial is rather tricky. The key point is the identity

$$a^3 + b^3 + c^3 - 3abc = (a + b + c) \cdot \frac{(a-b)^2 + (b-c)^2 + (c-a)^2}{2},$$

which shows that $a + b + c$ and $a^3 + b^3 + c^3 - 3abc$ have the same sign for all real numbers a, b, c, not all of them equal. The numbers $x, \sqrt[3]{2}y$ and $\sqrt[3]{3}z$ are equal if and only if $x = y = z = 0$, since $\sqrt[3]{2}$ and $\sqrt[3]{3}$ are irrational. So unless $x = y = z = 0$, the numbers $x + \sqrt[3]{2}y + \sqrt[3]{3}z$ and $x^3 + 2y^3 + 3z^3 - 3\sqrt[3]{6}xyz$ have the same sign. Note that this is trivially true when $x = y = z = 0$. To get rid of $\sqrt[3]{6}$, observe that $a - b$ and $a^3 - b^3$ have the same sign for all a, b, because the map $x \to x^3$ is increasing. Thus $x^3 + 2y^3 + 3z^3 - 3\sqrt[3]{6}xyz$ has the same sign as $(x^3 + 2y^3 + 3z^3)^3 - 27 \cdot 6(xyz)^3$. We conclude that

$$f(x, y, z) = (x^3 + 2y^3 + 3z^3)^3 - 27 \cdot 6(xyz)^3$$

is a solution of the problem.

28. Prove that there are infinitely many odd numbers in the sequence $\lfloor n\sqrt{2} \rfloor + \lfloor n\sqrt{3} \rfloor$ (with $n \geq 1$).

Solution. Let
$$x_n = \lfloor n\sqrt{2} \rfloor + \lfloor n\sqrt{3} \rfloor = y_n + z_n$$
with $y_n = \lfloor n\sqrt{2} \rfloor$ and $z_n = \lfloor n\sqrt{3} \rfloor$. Suppose that x_n is even for all $n \geq N$, where N is a certain positive integer. Then y_n and z_n have the same parity for $n \geq N$. On the other hand,
$$y_{n+1} - y_n = \lfloor n\sqrt{2} + \sqrt{2} \rfloor - \lfloor n\sqrt{2} \rfloor, \quad z_{n+1} - z_n = \lfloor n\sqrt{3} + \sqrt{3} \rfloor - \lfloor n\sqrt{3} \rfloor$$
are each either 1 or 2 (this follows from $\sqrt{2}, \sqrt{3} \in (1,2)$). Since y_n and z_n, as well as y_{n+1} and z_{n+1} have the same parity, it follows that $y_{n+1} - y_n$ and $z_{n+1} - z_n$ have the same parity and thus they must be equal for $n \geq N$. This means that
$$y_{n+1} - z_{n+1} = y_n - z_n$$
for $n \geq N$, thus the sequence $y_n - z_n$ becomes constant from a certain point on. This is certainly impossible, since
$$y_n - z_n < n\sqrt{2} - n\sqrt{3} + 1 = n(\sqrt{2} - \sqrt{3}) + 1$$
and the last quantity is smaller than any given number (in particular than $y_N - z_N$) for n large enough. This contradiction shows that the initial assumption is wrong and the result follows.

29. Prove that if $x_1, ..., x_n > 0$ satisfy $x_1 x_2 ... x_n = 1$, then
$$\left(\frac{x_1 + x_2 + ... + x_n}{n} \right)^{2n} \geq \frac{x_1^2 + x_2^2 + ... + x_n^2}{n}.$$

Solution. The key point here is the identity
$$x_1^2 + x_2^2 + ... + x_n^2 = (x_1 + x_2 + ... + x_n)^2 - 2 \sum_{i<j} x_i x_j.$$

The AM-GM inequality and the hypothesis $x_1 x_2 ... x_n = 1$ show that
$$\sum_{i<j} x_i x_j \geq \binom{n}{2}.$$

Hence
$$x_1^2 + x_2^2 + ... + x_n^2 \leq (x_1 + x_2 + .. + x_n)^2 - (n^2 - n).$$

It suffices therefore to prove that

$$\left(\frac{x_1 + x_2 + \ldots + x_n}{n}\right)^{2n} \geq \frac{(x_1 + x_2 + \ldots + x_n)^2 - (n^2 - n)}{n}.$$

The advantage is that this is no longer an inequality involving n variables, but only one, namely $S = x_1 + x_2 + \ldots + x_n$. Note that $S \geq n$, again by the AM-GM inequality and the hypothesis $x_1 x_2 \ldots x_n = 1$. So we need to prove that for all $S \geq n$ we have

$$\left(\frac{S}{n}\right)^{2n} \geq \frac{S^2}{n} - (n-1).$$

This is actually true for all $S > 0$, as the following easy application of the AM-GM inequality shows:

$$\left(\frac{S}{n}\right)^{2n} + n - 1 = \left(\frac{S}{n}\right)^{2n} + 1 + \ldots + 1 \geq n\sqrt[n]{\left(\frac{S}{n}\right)^{2n}} = \frac{S^2}{n}.$$

30. The equation $x^3 + x^2 - 2x - 1 = 0$ has three real solutions x_1, x_2, x_3. Find $\sqrt[3]{x_1} + \sqrt[3]{x_2} + \sqrt[3]{x_3}$.

Solution. Let $a = \sqrt[3]{x_1} + \sqrt[3]{x_2} + \sqrt[3]{x_3}$ and let

$$b = \sqrt[3]{x_1 x_2} + \sqrt[3]{x_2 x_3} + \sqrt[3]{x_3 x_1}.$$

The identity

$$u^3 + v^3 + w^3 - 3uvw = (u+v+w)((u+v+w)^2 - 3(uv+vw+wu))$$

yields

$$x_1 + x_2 + x_3 - 3\sqrt[3]{x_1 x_2 x_3} = a(a^2 - 3b)$$

and

$$x_1 x_2 + x_2 x_3 + x_3 x_1 - 3\sqrt[3]{(x_1 x_2 x_3)^2} = b(b^2 - 3\sqrt[3]{x_1 x_2 x_3} a).$$

Using Vieta's relations, the previous relations can be written as

$$-4 = a(a^2 - 3b) \quad \text{and} \quad -5 = b(b^2 - 3a).$$

Thus $a^3 = 3ab - 4$ and $b^3 = 3ab - 5$. Multiplying these relations, we obtain

$$(ab)^3 = 9(ab)^2 - 27ab + 20 \quad \text{or} \quad (ab-3)^3 = -7.$$

We conclude that $ab = 3 - \sqrt[3]{7}$ and finally

$$a = \sqrt[3]{3ab - 4} = \sqrt[3]{5 - 3\sqrt[3]{7}}.$$

Solutions to advanced problems 179

31. Prove that for all $a, b, c, x, y, z \geq 0$

$$(a^2 + x^2)(b^2 + y^2)(c^2 + z^2) \geq (ayz + bzx + cxy - xyz)^2.$$

Solution. First, we will try to get rid of some of the (many) variables. Let us note that if $xyz = 0$, say for simplicity $x = 0$, then the inequality becomes

$$a^2(b^2 + y^2)(c^2 + z^2) \geq a^2 y^2 z^2$$

and this is clear, since $b^2 + y^2 \geq y^2$ and $c^2 + z^2 \geq z^2$. Hence we may assume that $xyz \neq 0$. We set

$$u = \frac{a}{x}, \quad v = \frac{b}{y}, \quad w = \frac{c}{z}.$$

Dividing the original inequality by $(xyz)^2$ yields the equivalent inequality

$$(u^2 + 1)(v^2 + 1)(w^2 + 1) \geq (u + v + w - 1)^2.$$

Expanding, we obtain

$$(uvw)^2 + (uv)^2 + (vw)^2 + (uw)^2 + u^2 + v^2 + w^2 + 1 \geq$$
$$u^2 + v^2 + w^2 + 2(uv + vw + wu) - 2(u + v + w) + 1$$

or equivalently

$$(uvw)^2 + (uv)^2 + u + v + (vw)^2 + v + w + (uw)^2 + u + w \geq 2(uv + vw + wu).$$

This follows from the inequalities $(uvw)^2 \geq 0$ and (using the AM-GM inequality)

$$(uv)^2 + u + v \geq 3\sqrt[3]{(uv)^3} = 3uv \geq 2uv.$$

Adding up these inequalities yields the desired result.

32. The sequence a_n is defined by $a_1 = \frac{1}{2}$ and

$$a_{n+1} = \frac{a_n^2}{a_n^2 - a_n + 1}$$

for $n \geq 1$. Prove that $a_1 + a_2 + ... + a_n < 1$ for all positive integers n.

Solution. Induction is certainly not going to help here, since $(a_1 + a_2 + ... + a_n)_n$ is an increasing sequence. We write the recurrence relation as

$$\frac{1}{a_{n+1}} = 1 - \frac{1}{a_n} + \frac{1}{a_n^2},$$

which strongly suggests considering the sequence $x_n = \frac{1}{a_n}$. We have
$$x_{n+1} = x_n^2 - x_n + 1,$$
thus
$$x_{n+1} - 1 = x_n(x_n - 1).$$
This implies
$$\frac{1}{x_n - 1} - \frac{1}{x_{n+1} - 1} = \frac{1}{x_n} = a_n,$$
which allows us to compute $a_1 + a_2 + ... + a_n$ using a telescopic sum:
$$a_1 + a_2 + ... + a_n = \sum_{k=1}^{n} \left(\frac{1}{x_k - 1} - \frac{1}{x_{k+1} - 1} \right) = \frac{1}{x_1 - 1} - \frac{1}{x_{n+1} - 1}.$$

By hypothesis $x_1 = 2$, hence in order to conclude it suffices to see that $x_{n+1} > 1$. For this, it suffices to prove that $x_n > 1$ for all n, by induction on n. This is immediate, since it holds for $n = 1$ and if it holds for n, then
$$x_{n+1} - 1 = x_n(x_n - 1) > 0,$$
thus $x_{n+1} > 1$. The problem is solved.

33. Prove that for all $a, b, c > 0$
$$\frac{a+b+c}{\sqrt[3]{abc}} + \frac{8abc}{(a+b)(b+c)(c+a)} \geq 4.$$

Solution. We are here in one of those nasty situations in which we would like very much to apply the AM-GM directly to each term, but this approach fails rather badly. Indeed, everything works fine for the first term, but there is no way we can bound the second term from below (except saying that it is nonnegative, which is not very helpful). Hence we need something smarter than that. We start by applying the AM-GM inequality to the denominator of the second fraction:
$$(a+b)(b+c)(c+a) \leq \left(\frac{a+b+b+c+c+a}{3} \right)^3 = \frac{8}{27}(a+b+c)^3.$$

Hence it suffices to prove that
$$\frac{a+b+c}{\sqrt[3]{abc}} + \frac{27abc}{(a+b+c)^3} \geq 4.$$

The good news is that instead of three variables, we only have one now: if we set
$$x = \frac{a+b+c}{\sqrt[3]{abc}},$$

Solutions to advanced problems 181

then
$$\frac{27abc}{(a+b+c)^3} = \frac{27}{x^3}.$$

Hence we need to prove that
$$x + \frac{27}{x^3} \geq 4.$$

This follows from another smart application of the AM-GM inequality
$$\frac{x}{3} + \frac{x}{3} + \frac{x}{3} + \frac{27}{x^3} \geq 4\sqrt[4]{\frac{x^3}{27} \cdot \frac{27}{x^3}} = 4.$$

34. Find all real numbers x such that
$$\frac{x^2}{x-1} + \sqrt{x-1} + \frac{\sqrt{x-1}}{x^2} = \frac{x-1}{x^2} + \frac{1}{\sqrt{x-1}} + \frac{x^2}{\sqrt{x-1}}.$$

Solution. The key point is to observe that the product of the numbers
$$a = \frac{x^2}{x-1}, \quad b = \sqrt{x-1}, \quad c = \frac{\sqrt{x-1}}{x^2}$$

equals 1, and that the equation can be written
$$a + b + c = \frac{1}{a} + \frac{1}{b} + \frac{1}{c}.$$

Since $abc = 1$, the previous relation becomes
$$a + b + c = ab + bc + ca.$$

Let $S = a + b + c$, then a, b, c are solutions of the equation
$$t^3 - St^2 + St - 1 = 0.$$

This equation factors quite nicely, since
$$t^3 - St^2 + St - 1 = (t-1)(t^2+t+1) - St(t-1) = (t-1)(t^2+t(1-S)+1).$$

Thus one of the numbers a, b, c equals 1. Conversely, if one of a, b, c equals 1, say $a = 1$, then
$$\frac{1}{a} + \frac{1}{b} + \frac{1}{c} = 1 + b + c = a + b + c.$$

Thus any solution of the initial equation is a solution of one of the equations
$$x^2 = x - 1, \quad \sqrt{x-1} = 1, \quad \sqrt{x-1} = x^2.$$

The first equation has no solution, since the discriminant is negative. The second equation has the unique solution $x = 2$. Finally, the last equation can be written as $x^4 = x - 1$. Now, $x > 1$, so $x^4 > x$, and we obtain $x - 1 = x^4 > x$, a contradiction. Hence the third equation has no solution and we conclude that the answer of the problem is $x = 2$.

35. Let $x > 30$ be a real number such that $\lfloor x \rfloor \cdot \lfloor x^2 \rfloor = \lfloor x^3 \rfloor$. Prove that $\{x\} < \frac{1}{2700}$.

    Solution. Write $x = n + a$, with $n = \lfloor x \rfloor$ and $a = \{x\}$. Note that $a \in [0, 1)$ and by hypothesis $n \geq 30$. On the other hand, the condition $\lfloor x \rfloor \cdot \lfloor x^2 \rfloor = \lfloor x^3 \rfloor$ can be written as

    $$n \lfloor n^2 + 2na + a^2 \rfloor = \lfloor n^3 + 3n^2 a + 3na^2 + a^3 \rfloor$$

    or, by canceling n^3, as

    $$n \lfloor 2na + a^2 \rfloor = \lfloor 3n^2 a + 3na^2 + a^3 \rfloor.$$

    In particular, we must have

    $$2n^2 a + na^2 \geq n \lfloor 2na + a^2 \rfloor > 3n^2 a + 3na^2 + a^3 - 1,$$

    forcing in particular $a < \frac{1}{n^2} \leq \frac{1}{900}$. This is weaker than what we are asked to prove, but not that far away. The key point is that

    $$2na + a^2 < \frac{2n}{n^2} + \frac{1}{n^4} < 1,$$

    so that $\lfloor 2na + a^2 \rfloor = 0$ and so the previous relation yields

    $$\lfloor 3n^2 a + 3na^2 + a^3 \rfloor = 0, \quad \text{so} \quad 3n^2 a + 3na^2 + a^3 < 1.$$

    In particular, we obtain

    $$a < \frac{1}{3n^2} \leq \frac{1}{2700}$$

    and we are done.

36. Prove that for all $x, y > 0$ we have $x^y + y^x > 1$.

    Solution. The result is plain if one of $x, y \geq 1$, so suppose that $x, y < 1$. The proof in this case is quite tricky and based on Bernoulli's inequality: $(1 + a)^b \leq 1 + ab$ if $a \geq -1$ and $0 < b < 1$. We deduce that

    $$x^{1-y} = (1 + x - 1)^{1-y} \leq 1 + (x - 1)(1 - y) = x + y - xy < x + y,$$

Solutions to advanced problems 183

which can be rewritten as
$$x^y > \frac{x}{x+y}.$$

In a similar way we obtain $y^x > \frac{y}{x+y}$. Adding these two inequalities yields the desired result.

37. Solve in real numbers the system
$$\begin{cases} x^3 + x(y-z)^2 = 2 \\ y^3 + y(z-x)^2 = 30 \\ z^3 + z(x-y)^2 = 16. \end{cases}$$

Solution. We start by writing the system as
$$\begin{cases} x(x^2+y^2+z^2) - 2xyz = 2 \\ y(x^2+y^2+z^2) - 2xyz = 30 \\ z(x^2+y^2+z^2) - 2xyz = 16. \end{cases}$$

Let us denote
$$x^2 + y^2 + z^2 = S, \quad xyz = P.$$

Then the system becomes
$$\begin{cases} x = \frac{2P+2}{S} \\ y = \frac{2P+30}{S} \\ z = \frac{2P+16}{S}. \end{cases}$$

Squaring each equation and adding the resulting relations yields
$$S = x^2 + y^2 + z^2 = \frac{(2P+2)^2 + (2P+30)^2 + (2P+16)^2}{S^2},$$

that is
$$S^3 = (2P+2)^2 + (2P+30)^2 + (2P+16)^2.$$

On the other hand, multiplying the equations of the previous system gives
$$P = xyz = \frac{(2P+2)(2P+30)(2P+16)}{S^3},$$

thus
$$(2P+2)(2P+16)(2P+30) = PS^3.$$

Combining the last two relations, we obtain

$$(2P+2)(2P+16)(2P+30) = P((2P+2)^2 + (2P+16)^2 + (2P+30)^2).$$

Dividing by 4 and then expanding brutally and simplifying the resulting expression, we obtain the equation

$$P^3 + 4P - 240 = 0.$$

Fortunately, this has the integer solution $P = 6$. This is the unique real solution, since the map $x \to x^3 + 4x$ is increasing on $\mathbf{R}$, thus injective. Going back to the previous relations, we obtain

$$S^3 = \frac{14 \cdot 28 \cdot 42}{6} = 14^3,$$

thus $S = 14$. Finally, going back to the system

$$\begin{cases} x = \frac{2P+2}{S} \\ y = \frac{2P+30}{S} \\ z = \frac{2P+16}{S}. \end{cases}$$

we obtain the unique solution $(x, y, z) = (1, 3, 2)$.

38. Let a, b, c, d be positive real numbers such that $a + b + c + d = 4$. Prove that

$$\frac{a^4}{(a+b)(a^2+b^2)} + \frac{b^4}{(b+c)(b^2+c^2)} + \frac{c^4}{(c+d)(c^2+d^2)} + \frac{d^4}{(d+a)(d^2+a^2)}$$

is equal to at least 1.

Solution. Let A denote the left hand-side of the inequality and let us denote the following expression by B

$$\frac{b^4}{(a+b)(a^2+b^2)} + \frac{c^4}{(b+c)(b^2+c^2)} + \frac{d^4}{(c+d)(c^2+d^2)} + \frac{a^4}{(a+d)(a^2+d^2)}.$$

The main ingredient in the proof is the equality $A = B$, which follows from

$$A - B = \sum \frac{a^4 - b^4}{(a+b)(a^2+b^2)} = \sum (a - b) = 0.$$

Indeed,

$$\frac{a^4 - b^4}{(a+b)(a^2+b^2)} = \frac{(a^2-b^2)(a^2+b^2)}{(a+b)(a^2+b^2)} = \frac{a^2-b^2}{a+b} = a - b.$$

Hence the desired inequality $A \geq 1$ is equivalent to $A + B \geq 2$. On the other hand,
$$A + B = \sum \frac{a^4 + b^4}{(a+b)(a^2 + b^2)}$$
and it suffices to prove that
$$\frac{a^4 + b^4}{(a+b)(a^2 + b^2)} \geq \frac{a+b}{4}.$$

This is equivalent to
$$a^4 + b^4 \geq \frac{(a+b)^2}{4} \cdot \frac{a^2 + b^2}{2}$$
and follows from the inequalities $(a+b)^2 \leq 2(a^2 + b^2)$ and $(a^2 + b^2)^2 \leq 2(a^4 + b^4)$.

39. Nonnegative numbers a, b, c, d, e, f add up to 6. Find the maximal value of
$$abc + bcd + cde + def + efa + fab.$$

Solution. We write
$$abc + bcd + cde + def + efa + fab$$
$$= abc + abf + cde + def + bcd + efa$$
$$= ab(c+f) + de(c+f) + bcd + efa$$
$$= (ab + de)(c+f) + bcd + efa$$
$$= [(a+d)(b+e) - ae - bd](c+f) + bcd + efa$$
$$= (a+d)(b+e)(c+f) - aec - aef - bcd - bdf + bcd + efa$$
$$= (a+d)(b+e)(c+f) - aec - bdf.$$

Next, we bound $aec + bdf$ from below by 0 and we bound $(a+d)(b+e)(c+f)$ using the AM-GM inequality:
$$(a+d)(b+e)(c+f) \leq \left(\frac{a+d+b+e+c+f}{3}\right)^3 = 8,$$

the last equality being a consequence of the hypothesis. We conclude that
$$abc + bcd + cde + def + efa + fab \leq 8.$$

We still have to see whether the value 8 can actually be attained. Looking at the equality cases, we need
$$aec = bdf = 0, \quad a + d = b + e = c + f = 2.$$

Simply take $a = b = c = 2$ and $d = e = f = 0$. Thus the answer of the problem is 8.

Here is a different approach to this problem. Look at what happens if we fix b, c, e, f, and $a+d$ and only allow variations which raise and lower a and d in compensating amounts. We can write the desired quantity as

$$abc + bcd + cde + def + efa + fab = (a+d)(bc+ef) + (fb)a + (ce)d.$$

Only the last two terms will vary. If $fb > ce$, then the maximum will occur when we raise a as much as possible so that $d = 0$. If $fb < ce$, the opposite occurs and the maximum is at $a = 0$. If $fb = ce$, then we cannot change the sum with this variation and we might as well assume $d = 0$. Thus by symmetry we may assume $d = 0$.

Similarly we may assume either b or e vanishes and either c or f vanishes. Thus up to symmetry, we reduce to the problem of maximizing abc subject to $a + b + c = 6$. But this is easily covered by the AM-GM inequality $abc \leq \left(\frac{a+b+c}{3}\right)^3 = 8$ with equality if and only if $a = b = c = 2$. Hence the maximum is 8.

40. Solve the following equation

$$\lfloor x \rfloor + \lfloor 2x \rfloor + \lfloor 4x \rfloor + \lfloor 8x \rfloor + \lfloor 16x \rfloor + \lfloor 32x \rfloor = 12345.$$

Solution. Let $n = \lfloor x \rfloor$, so that

$$\lfloor 2^k x \rfloor = \lfloor 2^k n + 2^k \{x\} \rfloor = 2^k n + \lfloor 2^k \{x\} \rfloor.$$

Thus the equation becomes

$$(1 + 2 + \ldots + 32)n + \sum_{k=1}^{5} \lfloor 2^k \{x\} \rfloor = 12345.$$

On the other hand,

$$1 + 2 + \ldots + 32 = 2^6 - 1 = 63$$

and

$$0 \leq \sum_{k=1}^{5} \lfloor 2^k \{x\} \rfloor \leq \sum_{k=1}^{5} (2^k - 1) = 1 + 3 + 7 + 15 + 31 = 57.$$

We conclude that

$$12345 - 63n \in [0, 57).$$

But this is impossible, since we can easily check that $12345 \equiv 60 \pmod{63}$. Hence the equation has no solution.

41. Let x, y, z be real numbers such that $x + y + z = 0$. Prove that
$$\frac{x(x+2)}{2x^2+1} + \frac{y(y+2)}{2y^2+1} + \frac{z(z+2)}{2z^2+1} \geq 0.$$

Solution. We start by multiplying everything by 2 and adding up 1 to each fraction. Since
$$2x(x+2) + 2x^2 + 1 = 4x^2 + 4x + 1 = (2x+1)^2,$$
the inequality is equivalent to
$$\frac{(2x+1)^2}{2x^2+1} + \frac{(2y+1)^2}{2y^2+1} + \frac{(2z+1)^2}{2z^2+1} \geq 3.$$

Now, we make all denominators equal by writing (using the fact that $x = -(y+z)$, thus $x^2 = (y+z)^2$)
$$2x^2 + 1 = \frac{4}{3}x^2 + \frac{2}{3}(y+z)^2 + 1 \leq \frac{4}{3}x^2 + \frac{4}{3}(y^2+z^2) + 1 =$$
$$\frac{4}{3}(x^2 + y^2 + z^2) + 1.$$

Thus
$$\frac{(2x+1)^2}{2x^2+1} + \frac{(2y+1)^2}{2y^2+1} + \frac{(2z+1)^2}{2z^2+1} \geq$$
$$\frac{(2x+1)^2 + (2y+1)^2 + (2z+1)^2}{\frac{4}{3}(x^2+y^2+z^2) + 1}.$$

It remains to prove that the last quantity is at least 3. Well, the good news is that the hypothesis yields
$$(2x+1)^2 + (2y+1)^2 + (2z+1)^2 = 4(x^2+y^2+z^2) + 4(x+y+z) + 3$$
$$= 4(x^2+y^2+z^2+3),$$

thus
$$\frac{(2x+1)^2 + (2y+1)^2 + (2z+1)^2}{\frac{4}{3}(x^2+y^2+z^2) + 1} = 3$$

and we are done.

42. Solve in real numbers the system
$$\begin{cases} \frac{1}{x} + \frac{1}{2y} = (x^2 + 3y^2)(3x^2 + y^2) \\ \frac{1}{x} - \frac{1}{2y} = 2(y^4 - x^4). \end{cases}$$

Solution. Adding and subtracting the equations, and expanding the parentheses gives the equivalent system

$$\begin{cases} \frac{2}{x} = x^4 + 10x^2y^2 + 5y^4 \\ \frac{1}{y} = 5x^4 + 10x^2y^2 + y^4. \end{cases}$$

and finally

$$\begin{cases} 2 = x^5 + 10x^3y^2 + 5xy^4 \\ 1 = 5x^4y + 10x^2y^3 + y^5. \end{cases}$$

Now, if we are lucky and/or inspired, we recognize the successive terms in the expansion of

$$(x+y)^5 = x^5 + 5x^4y + 10x^3y^2 + 10x^2y^3 + 5xy^4 + y^5$$

and

$$(x-y)^5 = x^5 - 5x^4y + 10x^3y^2 - 10x^2y^3 + 5xy^4 - y^5.$$

Hence the system is equivalent to

$$\begin{cases} (x+y)^5 = 3 \\ (x-y)^5 = 1. \end{cases}$$

We obtain therefore the unique solution

$$x = \frac{\sqrt[5]{3}+1}{2}, \quad y = \frac{\sqrt[5]{3}-1}{2}.$$

43. Find all triples (x, y, z) of positive real numbers satisfying simultaneously the inequalities $x + y + z - 2xyz \leq 1$ and

$$xy + yz + zx + \frac{1}{xyz} \leq 4.$$

Solution. Dividing the first inequality by $xyz > 0$, we get

$$\frac{1}{xy} + \frac{1}{yz} + \frac{1}{zx} - 2 \leq \frac{1}{xyz}.$$

Adding this to the second inequality and regrouping gives

$$\left(xy + \frac{1}{xy}\right) + \left(yz + \frac{1}{yz}\right) + \left(zx + \frac{1}{zx}\right) \leq 6.$$

Solutions to advanced problems 189

However it is easy to see that for $t > 0$, $t + \frac{1}{t} \geq 2$ with equality if and only if $t = 1$. Applying this for $t = xy$, yz, and zx we get

$$6 \leq \left(xy + \frac{1}{xy}\right) + \left(yz + \frac{1}{yz}\right) + \left(zx + \frac{1}{zx}\right) \leq 6.$$

Hence we must have equality throught and $xy = yz = zx = 1$. This in turn gives $x = y = z = 1$ as the only solution.

Let us give an alternative solution, which uses the (complicated) hypotheses and classical inequalities to get some control on xyz. Let $a = xyz$. Then by the AM-GM inequality and the hypothesis we have

$$4 \geq \frac{1}{xyz} + xy + yz + zx \geq 4\sqrt[4]{x^2y^2z^2 \cdot \frac{1}{xyz}} = 4\sqrt[4]{xyz},$$

thus $xyz \leq 1$. We will prove that $xyz = 1$. Assuming the contrary and using again the AM-GM inequality, we obtain

$$4 \geq \frac{1}{xyz} + xy + yz + zx \geq \frac{1}{xyz} + 3\sqrt[3]{(xyz)^2}.$$

Let $xyz = a^3$, then $a \leq 1$ and the previous inequality becomes

$$4 \geq \frac{1}{a^3} + 3a^2, \quad \text{or} \quad 4a^3 \geq 3a^5 + 1.$$

Writing it as $a^3 - 1 \geq 3a^5 - 3a^3$ and dividing by $a - 1$ (which is negative) yields

$$a^2 + a + 1 \leq 3a^3(a + 1).$$

Finally, the hypothesis combined with the AM-GM inequality yield

$$2a^3 + 1 = 2xyz + 1 \geq x + y + z \geq 3a,$$

which can be written as $2a^3 - 2a \geq a - 1$ and then $2a(a + 1) \leq 1$. We conclude that

$$a^2 + a + 1 \leq 3a^3(a + 1) = 3a^2 \cdot a(a + 1) \leq \frac{3}{2}a^2 < 2a^2,$$

hence $a^2 > a + 1 > 1$, a contradiction with $a \leq 1$.

Thus our initial assumption $xyz < 1$ was wrong and we have $xyz = 1$. Now the first condition becomes $x + y + z \leq 3$. Thus we must have equality in the AM-GM inequality

$$x + y + z \geq 3\sqrt[3]{xyz}$$

and $x = y = z$. Since $xyz = 1$, we finally obtain the unique solution of the problem $x = y = z = 1$.

44. Let a, b, c be positive real numbers and let $x = a + \frac{1}{b}$, $y = b + \frac{1}{c}$, $z = c + \frac{1}{a}$. Prove that

$$xy + yz + zx \geq 2(x + y + z).$$

Solution. This is fairly difficult. First, we use the pigeonhole principle to obtain that two of the numbers x, y, z are either in $(0, 2)$ or $[2, \infty)$. Say x and y are these two numbers. Then $(x - 2)(y - 2) \geq 0$, thus $xy + 4 \geq 2(x + y)$. Thus all we need to show is that $yz + zx \geq 2z + 4$, which can be written as

$$z(x + y - 2) \geq 4.$$

On the other hand,

$$z(x+y-2) = \left(c + \frac{1}{a}\right)\left(a + \frac{1}{c} + b + \frac{1}{b} - 2\right) \geq \left(c + \frac{1}{a}\right)\left(a + \frac{1}{c}\right),$$

since $b + \frac{1}{b} - 2 \geq 0$ (this follows from the AM-GM inequality). Thus it suffices to prove that

$$\left(c + \frac{1}{a}\right)\left(a + \frac{1}{c}\right) \geq 4.$$

This can be written as $(ac + 1)^2 \geq 4ac$, or equivalently as $(ac - 1)^2 \geq 0$, which is clearly true. Note that we could have also used the Cauchy-Schwarz inequality

$$\left(c + \frac{1}{a}\right)\left(a + \frac{1}{c}\right) = \left(c + \frac{1}{a}\right)\left(\frac{1}{c} + a\right) \geq \left(\sqrt{c \cdot \frac{1}{c}} + \sqrt{\frac{1}{a} \cdot a}\right)^2 = 4$$

or the AM-GM inequality

$$\left(c + \frac{1}{a}\right)\left(a + \frac{1}{c}\right) = ac + \frac{1}{ac} + \frac{a}{c} + \frac{c}{a} \geq 2 + 2 = 4.$$

45. Let a, b be nonzero real numbers, such that $\lfloor an + b \rfloor$ is an even integer for all positive integers n. Prove that a is an even integer.

Solution. Let $x_n = \lfloor an + b \rfloor$ and let $y_n = \{an + b\}$, so that $x_n + y_n = an + b$. The hypothesis implies that $x_{n+1} - x_n$ is an even integer for all n. On the other hand, we have

$$x_{n+1} - x_n = \lfloor x_n + y_n + \lfloor a \rfloor + \{a\} \rfloor - x_n = \lfloor y_n + \{a\} \rfloor + \lfloor a \rfloor.$$

Solutions to advanced problems

We will discuss two cases, according to the parity of $\lfloor a \rfloor$. If this number is even, then the previous discussion shows that $\lfloor y_n + \{a\} \rfloor$ is even. But since this number is 0 or 1, it must be zero. Thus $x_{n+1} - x_n = \lfloor a \rfloor$ for all n, which gives $x_n = x_0 + n\lfloor a \rfloor$ for all n. Thus

$$an + b < \lfloor b \rfloor + n\lfloor a \rfloor + 1$$

for all n, which can be written as $n\{a\} < 1 - \{b\}$ for all n. This clearly forces $\{a\} = 0$ and so a is an integer. Since we also assumed that $\lfloor a \rfloor$ is even, it follows that a is an even number.

Suppose now that $\lfloor a \rfloor$ is odd. Then the same arguments show that we must have $\lfloor y_n + \{a\} \rfloor = 1$ for all n, then $x_{n+1} = x_n + 1 + \lfloor a \rfloor$ for all n and $x_n = x_0 + n(1 + \lfloor a \rfloor)$. But this means that

$$an + b \geq \lfloor b \rfloor + n(1 + \lfloor a \rfloor) \quad \text{so} \quad n(1 - \{a\}) \leq \{b\}$$

for all n, which is clearly impossible as $1 - \{a\} > 0$. Hence this second case is actually impossible and the result follows.

46. Let a, b, c be positive real numbers. Find all real numbers x, y, z such that

$$\begin{cases} ax + by = (x - y)^2 \\ by + cz = (y - z)^2 \\ cz + ax = (z - x)^2. \end{cases}$$

Solution. We will first solve the system as a linear system in the variables ax, by and cz. For that, we add up the equations to obtain

$$ax + by + cz = \frac{(x - y)^2 + (y - z)^2 + (z - x)^2}{2}$$

and then we come back to each equation. For instance, the first equation becomes

$$\frac{(x - y)^2 + (y - z)^2 + (z - x)^2}{2} - cz = (x - y)^2,$$

that is

$$cz = \frac{(y - z)^2 + (z - x)^2 - (x - y)^2}{2}$$
$$= \frac{(y - z)^2 + (z - x)^2 - ((y - z) + (z - x))^2}{2}$$
$$= -(y - z)(z - x) = (z - y)(z - x).$$

Doing the same manipulations to each equation, we obtain the system

$$\begin{cases} ax = (x-y)(x-z) \\ by = (y-x)(y-z) \\ cz = (z-x)(z-y). \end{cases}$$

Next, we multiply the first equation by $-(y-z)$, the second one by $-(z-x)$ and the last one by $-(x-y)$, to get

$$\begin{cases} -x(y-z) = \frac{(x-y)(y-z)(z-x)}{a} \\ -y(z-x) = \frac{(x-y)(y-z)(z-x)}{b} \\ -z(x-y) = \frac{(x-y)(y-z)(z-x)}{c}. \end{cases}$$

We add up these equations, exploiting the fact that

$$x(y-z) + y(z-x) + z(x-y) = 0.$$

We obtain

$$(x-y)(y-z)(z-x)\left(\frac{1}{a} + \frac{1}{b} + \frac{1}{c}\right) = 0.$$

By hypothesis, the term $\frac{1}{a} + \frac{1}{b} + \frac{1}{c}$ is positive, in particular nonzero. Thus we must have $(x-y)(y-z)(z-x) = 0$. Suppose for instance that $x = y$. Then the first equation of the original system becomes $(a+b)x = 0$, thus $x = y = 0$ (since $a, b > 0$). The other equations are equivalent to $cz = z^2$, thus $z = 0$ or $z = c$. Thus the case $x = y$ gives two solutions $(0,0,0)$ and $(0,0,c)$. We similarly get the other solutions $(a,0,0)$ and $(0,b,0)$ of the system, by treating the cases $y = z$ and $z = x$.

Remark 15.1. The key step in this problem is obtaining the relations

$$ax = (x-y)(x-z), \quad by = (y-x)(y-z), \quad cz = (z-x)(z-y).$$

Once we obtain these relations, we can also argue as follows to finish the solution: suppose that no two of x, y, z are equal. If x is either the largest or smallest of the three, then the equality $ax = (x-y)(x-z)$ says that $x > 0$. However if x is in the middle, then it says $x < 0$. Since we can argue similarly for y and z, we conclude that the largest and smallest of the three are positive, but the middle one is negative. This is patently absurd, hence we have a contradiction. Thus two of x, y, z must be equal. If $x = y$, then we get $ax = by = 0$ and hence $x = y = 0$. Thus $cz = z^2$, hence either $z = 0$ or $z = c$. The other two cases are similar. Thus the solutions are $(x, y, z) = (0,0,0), (a,0,0), (0,b,0)$, and $(0,0,c)$.

Solutions to advanced problems

47. Positive real numbers a, b, c add up to 1. Prove that

$$(ab + bc + ca)\left(\frac{a}{b^2 + b} + \frac{b}{c^2 + c} + \frac{c}{a^2 + a}\right) \geq \frac{3}{4}.$$

Solution. We will use Hölder's inequality to kill the denominators:

$$(a(b+1) + b(c+1) + c(a+1))(ab + bc + ca)\left(\frac{a}{b^2 + b} + \frac{b}{c^2 + c} + \frac{c}{a^2 + a}\right)$$

$$\geq \left(\sqrt[3]{a(b+1) \cdot ab \cdot \frac{a}{b^2 + b}} + \sqrt[3]{b(c+1) \cdot bc \cdot \frac{b}{c^2 + c}}\right.$$

$$\left. + \sqrt[3]{c(a+1) \cdot ca \cdot \frac{c}{a^2 + a}}\right)^3.$$

The nasty expression on the right hand-side is simply $(a + b + c)^3 = 1$ after simplifications. We conclude that

$$(ab+bc+ca)\left(\frac{a}{b^2 + b} + \frac{b}{c^2 + c} + \frac{c}{a^2 + a}\right) \geq \frac{1}{a(b+1) + b(c+1) + c(a+1)}$$

and it remains to see that the last quantity is greater than or equal to $\frac{3}{4}$. This follows from

$$a(b+1) + b(c+1) + c(a+1) = a + b + c + ab + bc + ca \leq$$

$$1 + \frac{(a + b + c)^2}{3} = 1 + \frac{1}{3} = \frac{4}{3}.$$

48. A sequence of nonnegative real numbers $(a_n)_{n \geq 1}$ satisfies

$$|a_m - a_n| \geq \frac{1}{m + n}$$

for all distinct positive integers m, n. Prove that if a real number c is greater than a_n for all $n \geq 1$, then $c \geq 1$.

Solution. Fix some $n > 1$ and let $i_1, ..., i_n$ be a permutation of $1, 2, ..., n$ such that $a_{i_1} \leq a_{i_2} \leq ... \leq a_{i_n}$. Then by hypothesis we have

$$a_{i_j} - a_{i_{j-1}} \geq \frac{1}{i_j + i_{j-1}}$$

for $j = 2, ..., n$. Adding up these relations yields

$$c \geq a_{i_n} - a_{i_1} \geq \sum_{j=2}^{n} \frac{1}{i_j + i_{j-1}}.$$

We will now bound from below the expression in the right hand-side, using the AM-HM inequality:

$$\sum_{j=2}^{n} \frac{1}{i_j + i_{j-1}} \geq \frac{(n-1)^2}{i_2 + \ldots + i_n + i_1 + \ldots + i_{n-1}} > \frac{(n-1)^2}{2(i_1 + \ldots + i_n)}.$$

Since $i_1, \ldots, i_n$ is a permutation of $1, 2, \ldots, n$, we have

$$2(i_1 + \ldots + i_n) = 2(1 + 2 + \ldots + n) = n(n+1).$$

We conclude that

$$c > \frac{(n-1)^2}{n(n+1)}$$

for all $n > 1$. This can also be written as

$$n^2(c-1) + n(c+2) - 1 > 0$$

for all $n > 1$ and immediately implies that $c - 1 \geq 0$ (otherwise the left hand-side becomes negative for n large enough). Thus $c \geq 1$, which is what we wanted to prove.

49. Let a, b, c, d be real numbers such that $a + b + c + d = 0$ and $a^7 + b^7 + c^7 + d^7 = 0$. Find all possible values of $a(a+b)(a+c)(a+d)$.

    Solution. Let $x^4 + Ax^3 + Bx^2 + Cx + D$ be the polynomial with roots a, b, c, d. By Vieta's relations we have $A = 0$. Let $S_n = a^n + b^n + c^n + d^n$. By Newton's relations we obtain

    $$S_1 = 0, \quad S_2 = -2B, \quad S_3 = -3C$$

    and then

    $$S_4 = -BS_2 - CS_1 - 4D = 2B^2 - 4D, \quad S_5 = -BS_3 - CS_2 - DS_1 = 5BC,$$

    and

    $$S_7 = -BS_5 - CS_4 - DS_3 = -5B^2C - C(2B^2 - 4D) - D(-3C)$$
    $$= 7C(D - B^2).$$

    By hypothesis we have $S_7 = 0$, hence $C = 0$ or $D = B^2$.

    We will consider separately these two cases. Suppose first that $C = 0$. Then a, b, c, d are roots of the polynomial $x^4 + Bx^2 + D$. But this polynomial is even, so $-a$ is also a root. Thus $-a \in \{a, b, c, d\}$ and then clearly $a(a+b)(a+c)(a+d) = 0$.

Suppose now that $D = B^2$. Thus

$$abcd = D = (ab + bc + cd + da + ca + bd)^2 = \frac{1}{4}(a^2 + b^2 + c^2 + d^2)^2,$$

the last equality being a consequence of the equality

$$(a + b + c + d)^2 = a^2 + b^2 + c^2 + d^2 + 2(ab + bc + cd + da + ac + bd)$$

and the hypothesis $a + b + c + d = 0$. Hence

$$4abcd = (a^2 + b^2 + c^2 + d^2)^2 \geq (4\sqrt[4]{a^2b^2c^2d^2})^2 = 16|abcd|.$$

We conclude that $abcd = 0$ and then $a^2 + b^2 + c^2 + d^2 = 0$, thus $a = b = c = d = 0$ in this case.

The conclusion is now clear: 0 is the only value taken by $a(a+b)(a+c)(a+d)$.

50. Let a, b, c be real numbers such that $a^2 + b^2 + c^2 = 9$. Prove that

$$2(a + b + c) - abc \leq 10.$$

Solution. This is a very tricky application of the Cauchy-Schwarz inequality. Since the inequality is symmetric, we may order a, b, c such that $|c| = \max(|a|, |b|, |c|)$. Since $a^2 + b^2 + c^2 = 9$ and since c^2 is the greatest among a^2, b^2, c^2, we have $c^2 \geq 3$, thus $a^2 + b^2 = 9 - c^2 \leq 6$. Now, we apply the Cauchy-Schwarz inequality as follows:

$$(2(a+b+c) - abc)^2 = (2(a+b) + c(2-ab))^2 \leq (4 + (2-ab)^2)((a+b)^2 + c^2).$$

Now,

$$(a+b)^2 + c^2 = a^2 + b^2 + c^2 + 2ab = 9 + 2ab.$$

Let $x = ab$. It suffices to prove that

$$(4 + (2-x)^2)(9 + 2x) \leq 100.$$

Expanding the left hand-side yields

$$(4 + (2-x)^2)(9 + 2x) = (8 - 4x + x^2)(9 + 2x) = 72 - 20x + x^2 + 2x^3.$$

We need to prove that

$$2x^3 + x^2 - 20x - 28 \leq 0.$$

Luckily, the left hand-side vanishes at $x = -2$, so we can factor out $x+2$ and obtain

$$2x^3+x^2-20x-28 = 2x^3+4x^2-3x^2-6x-14(x+2) = (x+2)(2x^2-3x-14).$$

The term $2x^2 - 3x - 14$ still vanishes at $x = -2$, so factoring again $x+2$ yields the equivalent inequality

$$(x+2)^2(2x-7) \leq 0.$$

Thus, we need to prove that $x \leq \frac{7}{2}$. This follows from

$$|x| \leq \frac{a^2+b^2}{2} \leq 3.$$

51. A real number $x > 1$ has the property that $x^n\{x^n\} < \frac{1}{4}$ for all $n \geq 2013$. Prove that x is an integer.

    Solution. We start by computing

    $$\lfloor x^{2n} \rfloor = \lfloor (\lfloor x^n \rfloor + \{x^n\})^2 \rfloor = \lfloor x^n \rfloor^2 + \lfloor \{x^n\}^2 + 2\lfloor x^n \rfloor \cdot \{x^n\} \rfloor.$$

    Next, by hypothesis $\{x^n\} < \frac{1}{4x^n} < \frac{1}{4}$ for $n \geq 2013$, thus

    $$\{x^n\}^2 + 2\lfloor x^n \rfloor \cdot \{x^n\} < \frac{1}{16} + 2x^n\{x^n\} < \frac{1}{16} + \frac{1}{2} < 1.$$

    Combining the last two relations, we obtain $\lfloor x^{2n} \rfloor = \lfloor x^n \rfloor^2$ for $n \geq 2013$. Fix such n and observe that $\lfloor x^{2^k n} \rfloor = \lfloor x^n \rfloor^{2^k}$ for all $k \geq 1$ (by induction on k, using the previously established relation $\lfloor x^{2m} \rfloor = \lfloor x^m \rfloor^2$ for $m \geq 2013$). Consequently, for $n \geq 2013$ and $k \geq 1$ we have

    $$(x^n)^{2^k} < \lfloor x^n \rfloor^{2^k} + 1 \quad \text{or} \quad \left(\frac{x^n}{\lfloor x^n \rfloor}\right)^{2^k} < 1 + \frac{1}{\lfloor x^n \rfloor^{2^k}} < 2.$$

    If $x^n > \lfloor x^n \rfloor$, then choosing k large enough contradicts the previous inequality. Hence $x^n = \lfloor x^n \rfloor$ and this for all $n \geq 2013$. In particular x^{2013} and x^{2014} are integers. But then $x = \frac{x^{2014}}{x^{2013}}$ is a rational number. Write $x = \frac{p}{q}$ for some relatively prime positive integers p, q. Since x^{2013} is an integer, q^{2013} divides p^{2013}. But since q is relatively prime to p^{2013}, it follows that $q = 1$ and so x is an integer.

52. Is there a sequence $(a_n)_{n \geq 1}$ of real numbers such that $a_n \in [0, 4]$ for all n and

    $$|a_m - a_n| \geq \frac{1}{|m-n|}$$

for all distinct positive integers m, n?

Solution. The answer is positive. We will prove that the sequence defined by $a_n = 4\{n\sqrt{2}\}$ is a solution of the problem. It is clear that $a_n \in [0, 4]$ for all n. On the other hand,

$$|a_m - a_n| = 4|(m-n)\sqrt{2} - (\lfloor m\sqrt{2}\rfloor - \lfloor n\sqrt{2}\rfloor)|$$

and it suffices to prove that

$$\left|\sqrt{2} - \frac{p}{q}\right| \geq \frac{1}{4q^2}$$

for all integers p, q with $q \neq 0$ (the desired inequality is then obtained by setting $q = m - n$ and $p = \lfloor m\sqrt{2}\rfloor - \lfloor n\sqrt{2}\rfloor$). We may assume that $q > 0$. Note that the inequality is clear if $p < 0$, so suppose that $p \geq 0$. Let us argue by contradiction and assume that

$$\left|\sqrt{2} - \frac{p}{q}\right| < \frac{1}{4q^2}$$

It follows that

$$\frac{1}{4q^2} > \frac{|\sqrt{2}q - p|}{q} = \frac{|2q^2 - p^2|}{q(\sqrt{2}q + p)} \geq \frac{1}{q(\sqrt{2}q + p)}.$$

The last inequality follows from the fact that $2q^2 - p^2$ is a nonzero integer (since $\sqrt{2}$ is irrational). We conclude from this last inequality that $p \geq (4 - \sqrt{2})q > 2q$. But then, using the triangle inequality, we obtain

$$2 < \frac{p}{q} \leq \sqrt{2} + \left|\sqrt{2} - \frac{p}{q}\right| < \sqrt{2} + \frac{1}{4},$$

which is certainly impossible. This contradiction finishes the solution of the problem.

53. Let $a, b, c, d \in [\frac{1}{2}, 2]$ be such that $abcd = 1$. Find the maximal value of

$$\left(a + \frac{1}{b}\right)\left(b + \frac{1}{c}\right)\left(c + \frac{1}{d}\right)\left(d + \frac{1}{a}\right).$$

Solution. This problem is very hard. First, we will try to manipulate the given expression, reducing it to something simpler. For that, note that

$$\left(a + \frac{1}{b}\right)\left(c + \frac{1}{d}\right) = ac + \frac{1}{bd} + \frac{a}{d} + \frac{b}{c} = 2ac + \frac{a}{d} + \frac{b}{c},$$

since $(ac)(bd) = 1$. Similarly,
$$\left(b + \frac{1}{c}\right)\left(d + \frac{1}{a}\right) = 2bd + \frac{b}{a} + \frac{d}{c}.$$

All in all, we obtain
$$\left(a + \frac{1}{b}\right)\left(b + \frac{1}{c}\right)\left(c + \frac{1}{d}\right)\left(d + \frac{1}{a}\right)$$
$$= \left(2ac + \frac{a}{d} + \frac{c}{b}\right)\left(2bd + \frac{b}{a} + \frac{d}{c}\right)$$
$$= 4abcd + 2bc + 2ad + 2ab + \frac{b}{d} + \frac{a}{c} + 2cd + \frac{c}{a} + \frac{d}{b}$$
$$= \frac{a}{c} + \frac{c}{a} + \frac{b}{d} + \frac{d}{b} + 2(ab + bc + cd + da) + 4,$$

again since $abcd = 1$. Next, we complete squares, writing the previous expression as
$$\left(\sqrt{\frac{a}{c}} + \sqrt{\frac{c}{a}}\right)^2 + \left(\sqrt{\frac{b}{d}} + \sqrt{\frac{d}{b}}\right)^2 + 2(ab + bc + cd + da).$$

The miracle is that we can still complete squares, since
$$\left(\sqrt{\frac{a}{c}} + \sqrt{\frac{c}{a}}\right) \cdot \left(\sqrt{\frac{b}{d}} + \sqrt{\frac{d}{b}}\right) = ab + bc + cd + da.$$

Indeed, we have for instance
$$\sqrt{\frac{a}{c} \cdot \frac{b}{d}} = \sqrt{\frac{ab}{cd}} = ab,$$

because $(ab)(cd) = 1$. Thus we can write
$$\left(a + \frac{1}{b}\right)\left(b + \frac{1}{c}\right)\left(c + \frac{1}{d}\right)\left(d + \frac{1}{a}\right) = \left(\sqrt{\frac{a}{c}} + \sqrt{\frac{c}{a}} + \sqrt{\frac{b}{d}} + \sqrt{\frac{d}{b}}\right)^2,$$

a rather miraculous identity.

Once we obtain this, we are in good shape, since we can start exploiting the fact that $a, b, c, d \in [1/2, 2]$. This implies that $\frac{a}{c} \in [1/4, 4]$, hence
$$\sqrt{\frac{a}{c}} + \sqrt{\frac{c}{a}} \leq 2 + \frac{1}{2} = \frac{5}{2}.$$

Similarly, we have
$$\sqrt{\frac{b}{d}} + \sqrt{\frac{d}{b}} \leq \frac{5}{2}$$
and so
$$\left(a + \frac{1}{b}\right)\left(b + \frac{1}{c}\right)\left(c + \frac{1}{d}\right)\left(d + \frac{1}{a}\right) \leq 25.$$

It remains to see whether this bound is actually attained. This is quite easy: take $a = 2, c = \frac{1}{2}$, $b = 2$ and $d = \frac{1}{2}$. Thus the answer of the problem is 25.

16 Other Books from XYZ Press

1. Andreescu, T.; Kane, J., *Purple Comet Math Meet! - the first ten years*, 2013.

2. Andreescu, T.; Dospinescu, G., *Straight from the Book*, 2012.

3. Andreescu, T.; Boreico, I.; Mushkarov, O.; Nikolov, N., *Topics in Functional Equations*, 2012.

4. Andreescu, T., *Mathematical Reflections - the next two years*, 2012.

5. Andreescu, T., *Mathematical Reflections - the first two years*, 2011.

6. Andreescu, T.; Dospinescu, G., *Problems from the Book*, 2008.

Printed by "Combinatul Poligrafic"
Com. nr. 90042